T0195487

How Have Deployments During the War on Terrorism Affected Reenlistment?

James Hosek, Francisco Martorell

Prepared for the Office of the Secretary of Defense

Approved for public release; distribution unlimited

 NATIONAL DEFENSE RESEARCH INSTITUTE

The research described in this report was prepared for the Office of the Secretary of Defense (OSD). The research was conducted in the RAND National Defense Research Institute, a federally funded research and development center sponsored by the OSD, the Joint Staff, the Unified Combatant Commands, the Department of the Navy, the Marine Corps, the defense agencies, and the defense Intelligence Community under Contract W74V8H-06-C-0002.

Library of Congress Cataloging-in-Publication Data

Hosek, James R.
How have deployments during the war on terrorism affected reenlistment? / James Hosek, Francisco Martorell.
p. cm.
Includes bibliographical references.
ISBN 978-0-8330-4733-5 (pbk. : alk. paper)
1. United States—Armed Forces—Recruiting, enlistment, etc. 2. United States—Armed Forces—Foreign service. 3. United States—Armed Forces—Operational readiness. 4. War on Terrorism, 2001-—Manpower—United States. I. Martorell, Francisco. II. Title.

UB323.H6687 2009
355.2'23620973—dc22
2009028247

Published 2009 by the RAND Corporation
1776 Main Street, P.O. Box 2138, Santa Monica, CA 90407-2138
1200 South Hayes Street, Arlington, VA 22202-5050
4570 Fifth Avenue, Suite 600, Pittsburgh, PA 15213-2665
RAND URL: http://www.rand.org/
To order RAND documents or to obtain additional information, contact
Distribution Services: Telephone: (310) 451-7002;
Fax: (310) 451-6915; Email: order@rand.org

Preface

The military operations in Iraq and Afghanistan have been the United States' longest military engagements since the Vietnam War and the most severe test of the all-volunteer force, with the possible exception of the Gulf War in 1991. More than 1.5 million service members were deployed between 2002 and 2007, many of them more than once, and the fast pace of deployment has been felt throughout the military. Soldiers and marines have faced a steady cycle of predeployment training and exercises, deployment itself, and postdeployment reassignment and unit regeneration. Service members not on deployment are nonetheless busy planning and supporting military operations, caring for injured service members, and attending to recruiting, training, and other responsibilities at home and abroad. Many service members are married, and deployments have disrupted their family routines and created stress from separation and reintegration. At the same time, the long hours, tension, uncertainty, and violence of deployments have stressed the service members sent to fight.

Remarkably, despite the pressures from deployments on service members and their families, reenlistment rates have been stable since 2002. The purpose of this monograph is to enhance understanding of whether deployments affected service members' willingness to stay in the military, as the stress caused by deployments would suggest, and how it was that reenlistment held steady.

This monograph should be of interest to the defense manpower policy and research communities, including the branches of service, the Office of the Secretary of Defense, the U.S. Department of Veterans Affairs, Congress, and such agencies as the U.S. Government Accountability Office and the Congressional Budget Office, as well as researchers in the government, at research organizations, and in academia.

The research was sponsored by the Office of Secretary of Defense and conducted within the Forces and Resources Policy Center of the RAND National Defense Research Institute, a federally funded research and development center sponsored by the Office of the Secretary of Defense, the Joint Staff, the Unified Combatant Commands, the Department of the Navy, the Marine Corps, the defense agencies, and the defense Intelligence Community.

For inquiries about this monograph, contact James Hosek at James_Hosek@rand.org and Francisco Martorell at Francisco_Martorell@rand.org, and for more information on RAND's Forces and Resources Policy Center, contact the Director, James Hosek, by email at James_Hosek@rand.org; by phone at 310-393-0411, extension 7183; or by mail at the RAND Corporation, 1776 Main Street, P.O. Box 2138, Santa Monica, California, 90407-2138. More information about RAND is available at www.rand.org.

Contents

Preface iii

Figures vii

Tables ix

Summary xiii

Acknowledgments xvii

Abbreviations xix

CHAPTER ONE

Introduction 1

CHAPTER TWO

Background and Review of Selected Literature 3

Background 3

Literature Review 12

    Deployment and Retention 12

    Mental Health 14

    Family Support 17

CHAPTER THREE

Modeling Deployment and Reenlistment 19

Utility Model of Deployment 19

CHAPTER FOUR

Data Sources and Analysis Samples 23

Data Sources 23

    Proxy PERSTEMPO File 23

    Joint Uniform Military Pay System (JUMPS) File 24

    Status of Forces Surveys and Linked Survey-Administrative Record Dataset 24

Dataset of Reenlistment Decisions 24

    Identifying Reenlistment Decisions 24

    Creation of the Analysis File 25

    Deployment Measures 26

    Reenlistment Bonuses 27

    Covariates and Key Subgroups 27

Survey-Administrative Linked Dataset 28

CHAPTER FIVE
Econometric Model 29

CHAPTER SIX
Empirical Results Using Survey Data 33
Baseline Estimates 33
Estimates Controlling for Overtime Work and Deviations from Expected Time Away from
    Home 35
Estimates Using Similar Specification to That Used in the Administrative Data Analysis 37
Conclusion 38

CHAPTER SEVEN
Empirical Results Using Administrative Data 39
Estimates of Deployment and Bonus Effects, 2002–2007 39
    Deployment Effects by Service and First or Second Term 39
    Estimates of the Effect of Selective Reenlistment Bonuses on Retention 42
Deployment Effects on Retention Over Time 43
    Did the Estimated Bonus Effect Also Change for the Army? 45
    Is the Army's Negative Deployment Effect in 2006–2007 the Result of Stop-Loss? 46
    Why Did the Deployment Effects Become Negative Only for the Army? 48
Hostile Deployment Effects by Extent of Deployment 48
Effects by Service Member Subgroups 54
Conclusions 58

CHAPTER EIGHT
The Role of Reenlistment Bonuses in Sustaining Retention 59
Illustrative Examples 59
Comparison to Deployment Pay and Potential Bonus Amounts 60
Reenlistment Bonus Prevalence and Generosity 63
Accounting for the Impact of Bonuses on Army First-Term Reenlistment 65
Conclusions 68

CHAPTER NINE
Conclusion 69

APPENDIXES
A. **A Model of Reenlistment Bonus Setting** 73
B. **Relationship Between Bias in Estimated Bonus Effect and Estimated Deployment**
   Effect 83
C. **Additional Regression Results** 85
D. **Comparison with Hansen and Wenger's Navy Pay Elasticity** 145

References 149

Figures

2.1. Active-Duty Personnel Receiving Hostile-Fire Pay, by Service.............................. 4
2.2. Percentage with Hostile Deployment in Three Years Prior to Reenlistment, Army First Term 5
2.3. Percentage with Hostile Deployment in Three Years Prior to Reenlistment, by Service, First Term 5
2.4. Percentage with Hostile Deployment in Three Years Prior to Reenlistment, by Service, Second Term.. 6
2.5. Prevalence of Hostile Deployment, Army 7
2.6. Casualties and Deaths per 1,000 Soldiers and Marines Receiving Hostile-Fire Pay 7
2.7. Average Months of Hostile Deployment in 36 Months Preceding Reenlistment Decision, Army.. 9
2.8. Average Number of Hostile Deployments in 36 Months Preceding Reenlistment Decision, Army.. 9
2.9. Distribution of Months Deployed in 36 Months Preceding Reenlistment Decision, by Service, First Term 10
2.10. Reenlistment Rate, by Service, First Term.. 11
2.11. Reenlistment Rate, by Service, Second Term 11
3.1. Utility and Hostile Deployment Time... 21
7.1. Estimated Effect of Hostile Deployment on Reenlistment for Hostile Deployment in 12 Months Prior to Reenlistment Decision, by Year of Decision and Service 44
7.2. Hostile Deployment Effect by Months Deployed in 36 Months Preceding Reenlistment Decision, Army and Marine Corps 50
7.3. Estimated Effects of Hostile Deployment in Year Prior to Decision and in First Two Years of Window on Reenlistment, by Year of Decision, Army and Marine Corps 51
7.4. Estimated Effects of Number of Hostile Deployments in Three Years Prior to Decision, Army and Marine Corps.. 53
7.5. Estimated Effects of Hostile Deployment in Year Prior to Decision, by Gender, Army First Term 55
7.6. Estimated Effects of Hostile Deployment in Year Prior to Decision, by Gender, Army Second Term 55
7.7. Estimated Effects of Hostile Deployment in Year Prior to Decision, by Whether Member Serves in Combat-Arms Occupation, Army First Term.......................... 56
7.8. Estimated Effects of Hostile Deployment in Year Prior to Decision, by Whether Member Serves in Combat-Arms Occupation, Army Second Term 56
7.9. Estimated Effects of Hostile Deployment in Year Prior to Decision, by Marital Status, Army First Term 57
7.10. Estimated Effects of Hostile Deployment in Year Prior to Decision, by Marital Status, Army Second Term 57

8.1.   Reenlistment Bonus Prevalence and Average Step, Army First Term 64
8.2.   Reenlistment Bonus Prevalence and Average Step, Marine Corps First Term 64
8.3.   Reenlistment Bonus Prevalence and Average Step, Navy First Term....................... 65
8.4.   Reenlistment Bonus Prevalence and Average Step, Air Force First Term 65
9.1.   Average Months of Hostile Deployment Over 36 Months Preceding First-Term
        Reenlistment Decision, Army and Marine Corps 71
9.2.   Average Number of Hostile Deployments Over 36 Months Preceding First-Term
        Reenlistment Decision, Army and Marine Corps 71

Tables

2.1. Prevalence of Exposure to Combat 16
2.2. Prevalence of Mental Health Conditions Following Deployment to OIF 17
6.1. Estimated Hostile Deployment Effects on Work Stress, Personal Stress, Intention
to Reenlist, and Reenlistment, First-Term Survey Respondents 34
6.2. Estimated Hostile Deployment Effects on Work Stress, Personal Stress, Intention
to Reenlist, and Reenlistment, Second-Term-Plus Survey Respondents 34
6.3. Estimated Hostile Deployment Effects, Controlling for Overtime Work and
Time Away from Home, First-Term Survey Respondents 36
6.4. Estimated Hostile Deployment Effects, Controlling for Overtime Work and
Time Away from Home, Second-Term-Plus Survey Respondents 36
6.5. Estimated Hostile Deployment Effects, Similar Specification as for
Administrative Data, First-Term Survey Respondents 37
6.6. Estimated Hostile Deployment Effects, Similar Specification as for
Administrative Data, Second-Term-Plus Survey Respondents 38
7.1. Estimated Deployment Effects on Reenlistment, First-Term Decisions, 2002–2007 40
7.2. Estimated Deployment Effects on Reenlistment, Second-Term Decisions,
2002–2007 41
7.3. Rough Estimate of Percentage Constrained by Stop-Loss, by Service, 1996–2007 47
8.1. Deployment and Compensating Bonus, Given Initial Expectation of Hostile
Deployment One-Fourth of the Time with Certainty 60
8.2. After-Tax Regular Military Compensation and Hostile Deployment–Related Pay
for an E-4 in the Fourth Year of Service, Monthly, 1996–2008 61
8.3. Hostile-Deployment Pay and Selective Reenlistment Bonus Relative to After-Tax
Regular Military Compensation for an E-4 in the Fourth Year of Service, Monthly,
1996–2008 62
8.4. Net Impact of Hostile Deployment and Bonus on Army First-Term Reenlistment,
by Year 66
8.5. Net Impact of Hostile Deployment and Bonus on Army First-Term Reenlistment,
Between Years 67
C.1. Glossary of Variables 86
C.2. Effect of Deployment on Work Stress, First-Term Survey Respondents 87
C.3. Effect of Deployment on Work Stress, Second-Term-Plus Survey Respondents 88
C.4. Effect of Deployment on Personal Stress, First-Term Survey Respondents 89
C.5. Effect of Deployment on Personal Stress, Second-Term-Plus Survey Respondents 90
C.6. Effect of Deployment on Intention to Reenlist, First-Term Survey Respondents 91
C.7. Effect of Deployment on Intention to Reenlist, Second-Term-Plus Survey
Respondents 92
C.8. Effect of Deployment on Reenlistment, First-Term Survey Respondents 93

C.9.   Effect of Deployment on Reenlistment, Second-Term-Plus Survey Respondents......... 94
C.10.   Effect of Deployment on Work Stress, First-Term Survey Respondents, with
        Controls for Overtime and Time Away More or Less than Expected......................95
C.11.   Effect of Deployment on Work Stress, Second-Term-Plus Survey Respondents,
        with Controls for Overtime and Time Away More or Less than Expected............... 96
C.12.   Effect of Deployment on Personal Stress, First-Term Survey Respondents, with
        Controls for Overtime and Time Away More or Less than Expected..................... 98
C.13.   Effect of Deployment on Personal Stress, Second-Term-Plus Survey Respondents,
        with Controls for Overtime and Time Away More or Less than Expected............... 99
C.14.   Effect of Deployment on Intention to Reenlist, First-Term Survey Respondents,
        with Controls for Overtime and Time Away More or Less than Expected.............. 101
C.15.   Effect of Deployment on Intention to Reenlist, Second-Term-Plus Survey
        Respondents, with Controls for Overtime and Time Away More or Less than
        Expected 102
C.16.   Effect of Deployment on Reenlistment, First-Term Survey Respondents, with
        Controls for Overtime and Time Away More or Less than Expected.................... 104
C.17.   Effect of Deployment on Reenlistment, Second-Term-Plus Survey Respondents,
        with Controls for Overtime and Time Away More or Less than Expected.............. 105
C.18.   Effect of Deployment on Reenlistment in Administrative Data, One-Year
        Window, First Term and No MOS Controls.. 107
C.19.   Effect of Deployment on Reenlistment in Administrative Data, One-Year
        Window, Second Term and No MOS Controls 108
C.20.   Effect of Deployment on Reenlistment in Administrative Data, One-Year
        Window, First Term and MOS Fixed Effects 109
C.21.   Effect of Deployment on Reenlistment in Administrative Data, One-Year
        Window, Second Term and MOS Fixed Effects.. 110
C.22.   Effect of Deployment on Reenlistment in Administrative Data, One-Year
        Window, First Term and MOS-by-Quarter Fixed Effects................................. 111
C.23.   Effect of Deployment on Reenlistment in Administrative Data, One-Year
        Window, Second Term and MOS-by-Quarter Fixed Effects.............................. 112
C.24.   Effect of Deployment on Reenlistment in Administrative Data, Three-Year
        Window, First Term and No MOS Controls.. 113
C.25.   Effect of Deployment on Reenlistment in Administrative Data, Three-Year
        Window, Second Term and No MOS Controls 114
C.26.   Effect of Deployment on Reenlistment in Administrative Data, Three-Year
        Window, First Term and MOS Fixed Effects 115
C.27.   Effect of Deployment on Reenlistment in Administrative Data, Three-Year
        Window, Second Term and MOS Fixed Effects.. 116
C.28.   Effect of Deployment on Reenlistment in Administrative Data, Three-Year
        Window, First Term and MOS-by-Quarter Fixed Effects................................. 117
C.29.   Effect of Deployment on Reenlistment in Administrative Data, Three-Year
        Window, Second Term and MOS-by-Quarter Fixed Effects.............................. 118
C.30.   Effect of Deployment in Prior Year on Reenlistment in Administrative Data,
        by Year of Decision 119
C.31.   Effect of Deployments in Prior Year and First Two Years of Three-Year Window,
        by Year of Decision, Army... 120
C.32.   Effect of Deployments in Prior Year and First Two Years of Three-Year Window,
        by Year of Decision, Navy 121

C.33. Effect of Deployments in Prior Year and First Two Years of Three-Year Window,
       by Year of Decision, Marine Corps... 122
C.34. Effect of Deployments in Prior Year and First Two Years of Three-Year Window,
       by Year of Decision, Air Force... 123
C.35. Effect of Months of Deployment in Prior Three Years on Reenlistment, by Year of
       Decision, Army.. 124
C.36. Effect of Months of Deployment in Prior Three Years on Reenlistment, by Year of
       Decision, Navy.. 125
C.37. Effect of Months of Deployment in Prior Three Years on Reenlistment, by Year of
       Decision, Marine Corps.. 126
C.38. Effect of Months of Deployment in Prior Three Years on Reenlistment, by Year of
       Decision, Air Force... 127
C.39. Effect of Number of Deployments in Prior Three Years on Reenlistment, by Year
       of Decision, Army... 128
C.40. Effect of Number of Deployments in Prior Three Years on Reenlistment, by Year
       of Decision, Navy... 129
C.41. Effect of Number of Deployments in Prior Three Years on Reenlistment, by Year
       of Decision, Marine Corps 130
C.42. Effect of Number of Deployments in Prior Three Years on Reenlistment, by Year
       of Decision, Air Force.. 131
C.43. Effect of Number of Deployments in Prior Year on Reenlistment, by Gender
       and Year of Decision, First Term... 132
C.44. Effect of Number of Deployments in Prior Year on Reenlistment, by Gender
       and Year of Decision, Second Term 133
C.45. Effect of Number of Deployments in Prior Year on Reenlistment, by Combat
       Arms and Year of Decision, First Term 134
C.46. Effect of Number of Deployments in Prior Year on Reenlistment, by Combat
       Arms and Year of Decision, Second Term 135
C.47. Effect of Number of Deployments in Prior Year on Reenlistment, by Marital
       Status and Year of Decision, First Term... 136
C.48. Effect of Number of Deployments in Prior Year on Reenlistment, by Marital
       Status and Year of Decision, Second Term 137
C.49. Effect of Reenlistment Bonus Multiplier and Deployment on Reenlistment,
       1996–2007 138
C.50. Descriptive Statistics for Administrative Data Sample, 1996–1997 139
C.51. Descriptive Statistics for Survey Sample, First-Term Respondents 141
C.52. Descriptive Statistics for Survey Sample, Second-Term-Plus Respondents 143

Summary

This monograph analyzes the relationship between deployment and reenlistment during the war on terrorism. It reviews trends in deployment and reenlistment and recent literature on deployment and its consequences. It also presents theoretical and econometric models of the effect of deployment on reenlistment and empirical estimates of the effect based on survey and administrative data. In addition, it presents estimates of the effect of bonuses on reenlistment and describes the role of bonuses in sustaining reenlistment.

Deployment Trends and Recent Literature

The key indicator of deployment in our analysis is the receipt of hostile-fire pay (HFP). The number of active-component service members receiving HFP was typically under 50,000 per month from 1996 to 2001, climbed rapidly to 300,000 per month by spring 2003, decreased to 150,000 per month by fall 2003, and climbed again to 200,000 per month in 2007. The increases were greatest for the Army and Marine Corps. Because the operations in Iraq and Afghanistan were staffed on a unit rotational basis, a large fraction of soldiers and marines were deployed for hostile duty. From 2003 to 2007, roughly 80 percent of soldiers at the first-term reenlistment and 60–80 percent of soldiers at the second-term reenlistment point had been deployed on hostile duty at some point during the three years prior to reenlistment. The literature review found generally positive effects of deployment on reenlistment but growing concern about the mental health consequences of deployment. Studies found that exposure to combat can have long-lasting effects on mental health, that exposure to combat was higher in Iraq and Afghanistan than in other locations, and that separation rates from the military for those with mental health concerns were higher for personnel who had been deployed to Iraq or Afghanistan as compared with other locations. However, a study found that, among service members who had married since 2002, the effect of deployment was to reduce the likelihood of marital dissolution.

Theoretical and Econometric Models

The theoretical model assumes that reenlistment depends on expected utility in the military versus the best alternative. Expected utility in the military depends on home time, deployed time, and income, including deployment pay. The model predicts that reenlistment will be lower if the amount of deployment time is much less or much greater than expected and that deployment pay and reenlistment bonuses can help to compensate for this. In joining the military, individuals will select the service and occupation most in line with their preference for deployment. Therefore, controls for service and occupation in empirical work will help to control for selection.

The econometric model relates reenlistment to deployment, bonuses, and other variables. We discuss possible biases in estimates of the deployment effect and the bonus effect, and we suggest that deployment effects are likely to be unbiased in models controlling for occupational specialty, service, and year or, alternatively, for occupation by quarter and service. Occupation, service, and year controls also help to reduce possible bias in the bonus effect, but not all of the bias will be eliminated, and the remaining bias may be either negative or positive.

Findings from the Survey Data

Our survey data consist of 10 Status of Forces Surveys of Active Duty Personnel administered via the Internet in 2002 to 2005. The surveys have been linked to administrative files on personnel and pay. The survey data contain variables not available in the administrative data, such as work stress, personal stress, intention to stay in the military, number of days longer than the usual duty day, whether time away was less or more than expected, and individual and unit preparedness. Similarly, the administrative data contain information not available in the survey, such as actual reenlistment, deployment, Armed Forces Qualification Test (AFQT) category, and reenlistment bonus. Regression analysis pooled the respondents from the various surveys. Deployment is coded according to deployment involving hostile duty and deployment involving no hostile duty ("nonhostile deployment only").

The key findings are that, among survey respondents, hostile deployment tended to increase work stress and personal stress and reduce the intention to reenlist. It also tended to reduce actual first-term reenlistment but increase second-term-plus reenlistment. The effects of deployment change when variables for overtime work (the number of days longer than the usual duty day) and whether the respondent spent more or less time away than expected are added to the model. With those variables included, first-term hostile deployment had no effect or a negative effect on work stress and personal stress, a negative effect on intention to reenlist, and no effect on actual reenlistment. The results are similar for second-term-plus personnel, except that hostile deployment has no effect on intention to reenlist (with the exception of the Navy, where the effect is positive). The effect on actual reenlistment is positive. The results suggest two reasons that deployment increases stress and reduces the intention to reenlist: Deployments often involve long days, and expectations about the length of deployment might not be met.

Models with a specification using variables available in the administrative data allow a comparison between the estimates for survey respondents versus all service members at reenlistment. In these models, the effect of hostile deployment on first-term reenlistment for survey respondents is negative in the Army, zero in the Navy and Marine Corps, and positive in the Air Force. By comparison, the effect of hostile deployment on first-term reenlistment for the population of service members at reenlistment is negative in the Army and Navy, positive in the Marine Corps, and zero in the Air Force. The effect of hostile deployment on second-term-plus reenlistment for survey respondents is zero in the Army and Air Force and positive in the Navy and Marine Corps. Although there are differences between estimates based on the survey data and those based on administrative data, there is no systematic pattern for first-term reenlistment such that, say, the effects of hostile deployment are lower in the survey estimates, which might reflect survey response bias. For second-term-plus reenlistment, however, the hostile deployment effects are lower for the survey respondents than for the population. Thus, the findings suggest possible survey response bias at second-term-plus reenlistment, with the survey respondents who were deployed being less likely to reenlist compared to the total population.

Findings from the Administrative Data

We estimate reenlistment models for first- and second-term reenlistment reenlistment for each service, and we explore many different specifications of the deployment variable, not all of which are summarized here. The models include controls for deployment that involved no hostile duty, years of service at the time of the decision, education, gender, AFQT, race, and an indicator for being promoted more rapidly than is typical. In addition, there are controls for occupational specialty, selective reenlistment bonus multiplier, and year of decision or, when bonus is omitted, for occupational specialty and quarter of decision.

An important advantage of the administrative data is a sufficient number of observations to analyze deployment effects by year. Using an indicator of hostile deployment in the year before reenlistment, we find the following deployment effects by year in 1996–2007:

- First term:
  - Army: positive but decreasing, turning negative in 2006 and 2007
  - Navy: remained near zero
  - Air Force: remained near zero
  - Marine Corps: remained near zero with upturn in 2006 and 2007
- Second term:
  - Army: positive but decreasing, turning negative in 2006 and 2007
  - Navy: positive, decreasing to zero in 2003, then increasing
  - Air Force: positive but decreasing to 2003, then increasing
  - Marine Corps: positive, decreasing to zero in 2003, then increasing.

The change from positive to negative in the effect of deployment on Army reenlistment is notable. Further analysis of the Army data indicates that the effect of deployment became negative because the effect of deployment on reenlistment was negative for those with a high number of months of deployment (12–17 months or 18 or more months) but positive for those with few months of deployment (1–6 months or 7–12 months). By 2006, two-thirds of the soldiers at reenlistment had been deployed for 12 or more months, and the soldiers in this category were subject to a negative deployment effect. Thus, a high cumulative number of months of deployment had a negative effect on reenlistment, and many soldiers had a high cumulative number of months of deployment. We find that it is unlikely that stop-loss caused the negative deployment effect in 2006–2007, though stop-loss might have added to the downward trend in deployment effect. These results are especially interesting when contrasted with those for the Marine Corps, which also experienced a marked increase in the fraction of personnel deployed and in mean months deployed. When broken out by months deployed, we actually find that a high number of months deployed (18 or more) also had a negative effect for marines at first-term reenlistment. But since these effects are not as large as they are for the Army, and since relatively few marines are in the high-deployment-month bins, the aggregate deployment effect for marines does not turn negative.

The deployment effects for men and women in the Army were similar, though the effect for men was more negative in 2006, probably because of a higher prevalence of men in combat arms. The downward trend in deployment effect was greater in combat arms than in non-combat arms, as might be expected given the greater likely exposure to combat in the combat-arms occupations. The effect of deployment on reenlistment was typically positive and higher for marrieds than for singles, though the effect became negative for both groups in 2006.

The Role of Reenlistment Bonuses

Despite the decreasing effect of deployment on reenlistment in the Army, the Army's reenlistment rate did not decline. Our analysis suggests that the Army's expanded use of bonuses and increased generosity of bonuses provided a positive impetus to reenlist that helped to offset the decreasing, then negative, effect of deployment on reenlistment.

Our estimates of the bonus effect on reenlistment are positive for all services but may be biased. Bias can arise from bonus-setting behavior in response to anticipated high or low reenlistment (reverse causality) or unobserved actions correlated with the bonus, such as career-counselor effort, other reenlistment incentives (such as choice of location or assignment), or limiting the number of reenlistment slots in certain occupations. Our methods partially control for these sources of bias. A reenlistment bonus experiment (i.e., a randomized controlled trial) may be necessary to eliminate bias.

Possible Importance of Deployment Episode Length

Soldiers and marines were extensively involved in the ground operations in Iraq, and the increase in average months of deployment was similar for both, though lower for marines. However, the effect of deployment on first-term reenlistment decreased over time for the Army and became negative in 2006, while the effect of deployment remained near zero for the Marine Corps. Army deployments were 12–15 months long, while Marine Corps deployments were seven months; marines had more episodes of deployment than did soldiers. It is worth studying whether longer deployments, and more prolonged exposure to combat, lead to lower reenlistment and higher prevalence of subsequent mental health conditions, and, if so, what might be done to avoid those outcomes.

Acknowledgments

We wish to thank our project sponsor, Jeanne Fites, Deputy Under Secretary of Defense for Program Integration (Personnel and Readiness), for her guidance and support throughout this project. We relied on data provided by the Defense Manpower Data Center. In particular, Timothy Elig helped us secure the Status of Forces Survey data and provided comments on an earlier draft, and Teri Cholar provided the Proxy PERSTEMPO dataset and patiently answered our questions.

Arthur Bullock at RAND oversaw data management and construction of the analytical files. His assistance was fundamental to the quality of our analysis. We also extend sincere thanks to our reviewers, Michael Hansen and Erik Meijer, whose comments led to substantial improvement of the monograph. Our editor, Lauren Skrabala, handled our long and often technical manuscript with outstanding skill.

Abbreviations

ACOL	annualized cost of leaving
AFQT	Armed Forces Qualification Test
CZTE	combat-zone tax exclusion
DMDC	Defense Manpower Data Center
DoD	U.S. Department of Defense
ETS	expiration of term of service
FY	fiscal year
GWOT	global war on terrorism
HFP	hostile-fire pay
IA	individual augmentation
IED	improvised explosive device
JUMPS	Joint Uniform Military Pay System
MOS	military occupational specialty
OEF	Operation Enduring Freedom
OIF	Operation Iraqi Freedom
PDHA	postdeployment health assessment
PERSTEMPO	personnel tempo
PTSD	post-traumatic stress disorder
RMC	regular military compensation
SRB	selective reenlistment bonus

CHAPTER ONE

Introduction

The military operations in Iraq and Afghanistan have required the largest and longest use of U.S. military forces since the Vietnam conflict. The buildup of forces for these operations began in 2002, and, by 2007, more than 1.5 million service members had been deployed. Army deployments have been about 12–15 months in length, Marine Corps deployments about seven months, Navy deployments (typically on board ship) about six months, and Air Force deployments three months or longer. Deployments are generally periods of high stress, when daily activities vary from humdrum routine to fast-paced action and traumatizing events, especially for ground forces. Stress is also high among support units at home, as both deployed and nondeployed service members in support activities often work long hours. To varying degrees, deployed service members have been able to stay in contact with family via phone calls and the Internet. In addition, close friendships can develop between "battle buddies" who rely on each other during missions and for social and psychological support. Departure for deployment means separation from friends and family, and the return from deployment results in an effort to reestablish those relationships. Married service members face the challenge of reintegrating with their families after the spouse has been responsible for running the household, and the children may be anxious and unsure of how to relate to their returning parent.

The deployments to Iraq and Afghanistan differ from other post-Vietnam deployments. Apart from the first Gulf War in 1991, which involved the deployments of nearly a million service members during a short period of fighting, deployments of U.S. forces have been directed to peacekeeping, peacemaking, and humanitarian purposes, e.g., Bosnia, Kosovo, Haiti, and Banda Aceh. In contrast, the defeat of Saddam's forces in spring 2003 was followed by insurrection and terrorism, with the tide turning in 2007 as U.S. forces surged and tribal leaders in western Iraq shifted their cooperation from al Qaeda to the United States. The Army and Marine Corps staffed the operations in Iraq and Afghanistan via unit rotation: Rather than permanently stationing service members in theater, units were rotated in and out. Many units, and many unit members, had more than one deployment. By fall 2008, about 4,700 service members had died in Operations Iraqi Freedom and Enduring Freedom, and more than 30,000 had been wounded in action (DMDC, 2008).

The first studies of the relationship between deployment and retention (Hosek and Totten, 1998, 2002; Fricker, 2002) found a positive relationship between deployment and retention regardless of whether deployment was measured in terms of total months deployed or the number of episodes of deployment. For most enlisted personnel and officers who were deployed, the first- and second-term reenlistment rate or the continuation rate (for junior officers) was higher than that of comparable personnel who were not deployed. However, there was some evidence that soldiers with many months of deployment or a large number of deployments during a 36-month period had a lower reenlistment rate than did soldiers who did not deploy. The data for these studies came from the 1990s, a time when deployments were typically not as long as deployments to Iraq and Afghanistan and when multiple deployments involving hostile duty were less frequent.

Our analysis extends the previous research in several ways. We consider the reenlistment of enlisted personnel from 1996 to 2007, whereas Hosek and Totten (2002) covered the period from 1996 to 1999. We contrast the relationship between deployment and retention before versus during the global war on terrorism (i.e., 1996–2001 versus 2002–2007), and we present estimates for data pooled across these years as well as for each year. We estimate models with a number of different empirical specifications for deployment as a check on the robustness of our findings, and we attempt to estimate the effect of reenlistment bonuses on reenlistment. We estimate models that pool data across occupational specialties and by combat arms and gender.

One of our chief findings is that the effect of deployment on Army reenlistment changed from a positive one prior to 2002 and during the first few years of the operations in Iraq and Afghanistan to a negative one in 2006 and 2007. The shift from positive to negative occurred despite significant deployment-related pay. Although the effect of deployment decreased from positive to negative, overall reenlistment rates remained fairly constant throughout 2002–2007 and were similar to pre-2002 rates. We think that an important reason for this was the Army's expanded use of reenlistment bonuses in 2005–2007.

The following chapters begin with descriptive information on deployments and a review of selected studies published since 2002 on deployment and retention, mental health and retention, and family support (Chapter Two). We then present a theoretical model of deployment and retention (Chapter Three) and describe our data and econometric approach (Chapters Four and Five). Chapters Six and Seven provide empirical results based on survey data that were linked to administrative records on reenlistment and on the full file of administrative data. Chapter Eight discusses the role of reenlistment bonuses in stabilizing reenlistment rates, and Chapter Nine presents our closing thoughts. Four appendixes describe the theoretical model in greater detail and catalog additional regression results that support the analysis presented here.

CHAPTER TWO

Background and Review of Selected Literature

Background

The buildup of personnel for Operation Iraqi Freedom (OIF) began in 2002 and peaked at 250,000 in spring 2003 as U.S. and coalition forces engaged and defeated Saddam Hussein's armed forces. The United States quickly reduced its forces to around 124,000 after the victory, and the objectives became stabilization, reconstruction, and humanitarian assistance. The months following Saddam's defeat did not secure the peace but, instead, saw the emergence of sectarianism as Shias gained political strength over the once-dominant Sunnis, the rise of insurgency as groups such as cleric Moktada al-Sadr's militia sought to dislodge U.S. forces, the growing use of improvised explosive devices (IEDs) that became increasingly sophisticated and lethal, and an increase in terrorist attacks and suicide bombings. Facing the prospect of a Balkanized, dysfunctional Iraq without a capable military, in 2007, the United States added more than 20,000 troops to its contingent in Iraq. By the end of 2007, violence in Iraq was down sharply (Campbell, O'Hanlon, and Unikewicz, 2007). Some observers have attributed the reduction in violence, at least in part, to the success of the "surge" strategy, along with the shift in allegiance of tribal chieftains in the western provinces, away from the insurgents toward the United States (Feldman, 2008). However, many aspects of Iraq's political stability and national security remained uncertain as 2008 approached (Filkins, 2008).

In Afghanistan, Operation Enduring Freedom (OEF) helped to dislodge al Qaeda and remove the Taliban from power in 2002, and a multinational coalition, the International Security Assistance Force, sought to stabilize the country. After a period of relative quiet, the Taliban reasserted itself with attacks that were initially sporadic but became frequent by 2006. The United States sustained a military presence in Afghanistan of about 18,000 troops from 2002 to 2008.

Our study covered the period from 1996 through 2007. We used receipt of hostile-fire pay (HFP) to indicate months in which a service member served in an area deemed hostile. Since 2002, the major hostile areas have been Iraq and Afghanistan, though in prior years, 40,000–50,000 active-duty service members received HFP each month. These personnel served in Bosnia or Kosovo, for example, or simply were on duty in dangerous areas.[1]

---

[1] The list of hostile-fire pay areas as of March 2008 included Afghanistan, Algeria, certain areas of the Arabian Peninsula and adjacent sea areas, Azerbaijan, Bahrain, Burundi, Chad, Colombia, Cote d'Ivoire, Cuba (Guantanamo), Democratic Republic of Congo, Djibouti, East Timor, Egypt, Eritrea, Ethiopia, certain areas of Greece, Haiti, Indonesia, Iran, Iraq, Israel, Jordan, Kenya, Kosovo, Kuwait, Kyrgyzstan, Lebanon, Liberia, Malaysia, Montenegro, Oman, Pakistan, Philippines, Qatar, Rwanda, Saudi Arabia, Serbia, Somalia, Sudan, Syria, Tajikistan, certain areas of Turkey, Uganda, United Arab Emirates, Uzbekistan, and Yemen (Office of the Under Secretary of Defense [Comptroller], 2009).

Figure 2.1 depicts the number of active-duty personnel by branch of service receiving HFP, by month, from 1996 to 2007.

Our data do not identify where a service member was stationed or what military operation he or she took part in, but the sharp increase in the receipt of HFP in 2003 no doubt resulted from the onset of OIF and OEF, which added to the existing level of operations. The United States might have reduced its staffing in other hostile areas and redeployed those personnel to Iraq and Afghanistan, so the increase in 2002–2007 relative to 1996–2001 is probably a conservative estimate of forces involved in OIF and OEF. As Figure 2.1 suggests, after Saddam's defeat in spring 2003, about 125,000–150,000 U.S. service members were deployed in any given month in OIF and OEF, or 11.3–13.6 percent of the 1.1 million personnel in the active-duty force. The Army and Marine Corps accounted for most of the increase in service members receiving HFP.

The number of personnel receiving HFP in a given month is not a good indicator of the number of service members who deployed to Iraq or Afghanistan, however, because OIF and OEF were staffed on a rotational basis. Under rotational staffing, many soldiers experienced deployment during their first term of service (see Figure 2.2). In 1996–2001, 10–20 percent of first-term soldiers had been on a hostile deployment in the 12 months preceding their reenlistment decision, and 20–30 percent had done so in the 36 months preceding their reenlistment decision. By the end of 2003, both of these percentages had soared to 70 percent, and in 2004–2007, 60 percent or more had been deployed in the preceding 12 months, and roughly 80 percent had been deployed in the preceding 36 months.

Figures 2.3 and 2.4 show the percentage of service members with hostile deployment in the 36 months preceding reenlistment, by branch of service for first- and second-term

Figure 2.1
Active-Duty Personnel Receiving Hostile-Fire Pay, by Service

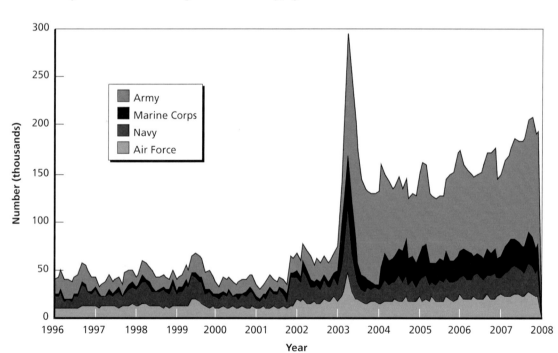

Figure 2.2
Percentage with Hostile Deployment in Three Years Prior to Reenlistment, Army First Term

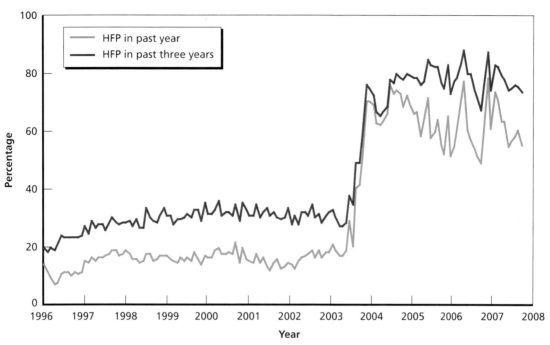

Figure 2.3
Percentage with Hostile Deployment in Three Years Prior to Reenlistment, by Service, First Term

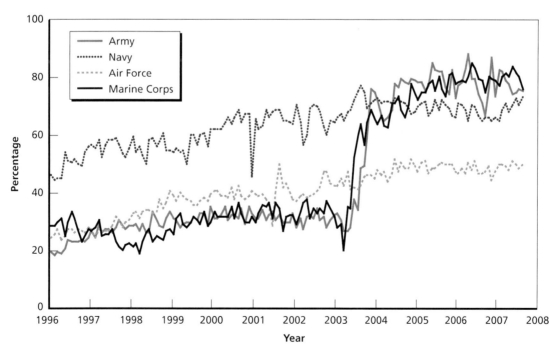

Figure 2.4
Percentage with Hostile Deployment in Three Years Prior to Reenlistment, by Service, Second Term

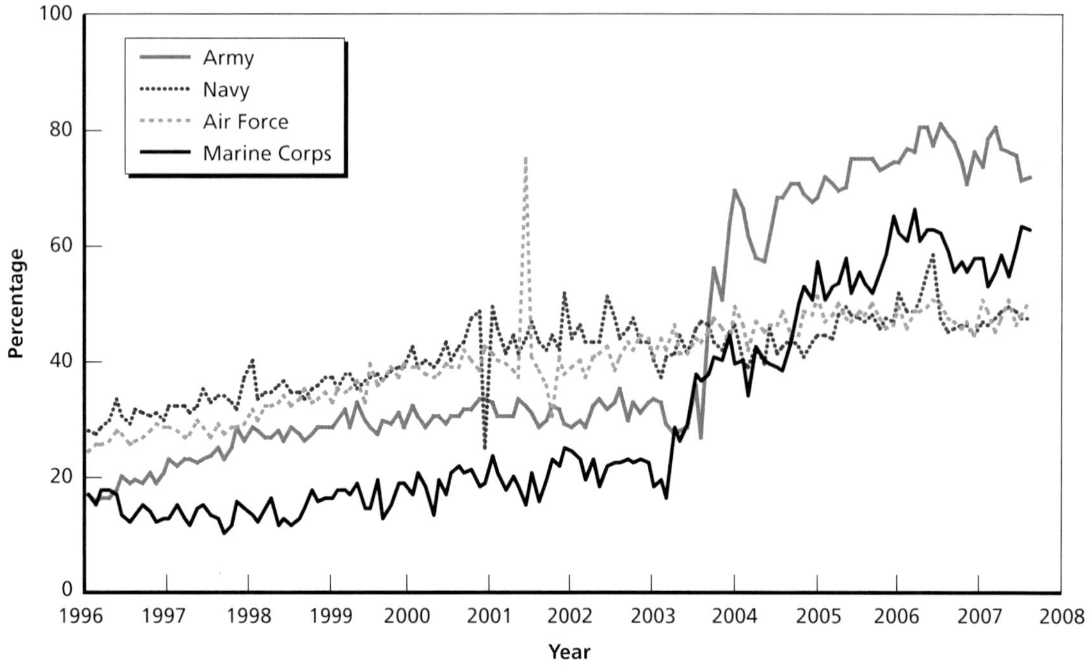

RAND *MG873-2.4*

personnel. The first-term Marine Corps experience was almost identical to that of the Army, with a doubling of the percentage with hostile deployment in 2003–2004. The Navy and Air Force showed slightly higher deployment in 2002–2007 versus 1996–2001. The patterns for second-term reenlistment (see Figure 2.4) were the same as those for the first term for the Army and Air Force. But in the Navy, the percentage with deployment at second-term reenlistment was lower (e.g., 40–50 percent versus 60–70 percent at first term), and the Marine Corps' increase in deployment in 2003–2004 was lower than in the first term.

Deployment varied by military occupational specialty (see Figure 2.5). The percentage of Army personnel with deployment in the previous three years in 2002–2007 averaged above 60 percent in many occupational areas versus about half that level in 1996–2001. These included combat-arms personnel (infantry, gun crews), electronic equipment repairers, communication and intelligence specialists, electrical/mechanical equipment repairers, craftsworkers, and service and supply handlers. The 2002–2007 rates were about 60 percent for other technical and allied specialists and 50 percent for health care specialists and functional support and administration personnel—and were less than half that in 1996–2001. The pattern for marines (not shown) was similar.

From 2002 to 2007, the vast majority of service-member casualties and deaths occurred in Iraq rather than Afghanistan. To gauge the danger facing troops on the ground in Iraq, we formed the ratio of the sum of those killed or wounded in action in Iraq per 1,000 receiving HFP.[2] Figure 2.6 displays the ratios for the Army and Marine Corps. The ratios tend to move

---

[2] Data on the number of service members killed and wounded in action are from the Defense Manpower Data Center (DMDC, 2008). Data on those receiving HFP are from our database (described in Chapter Three) and do not distinguish where HFP was received. Therefore, our denominator includes HFP recipients in other places in addition to Iraq.

**Figure 2.5
Prevalence of Hostile Deployment, Army**

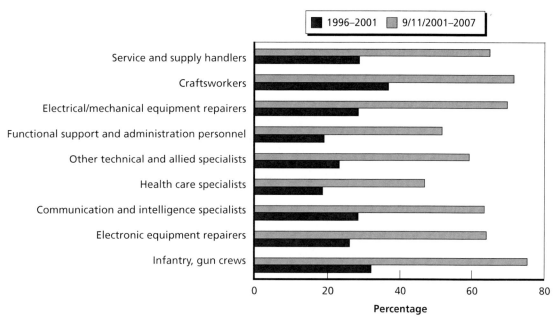

NOTE: Prevalence is defined as the percentage with a hostile deployment in the 36 months prior to the reenlistment decision.

RAND *MG873-2.5*

**Figure 2.6
Casualties and Deaths per 1,000 Soldiers and Marines Receiving Hostile-Fire Pay**

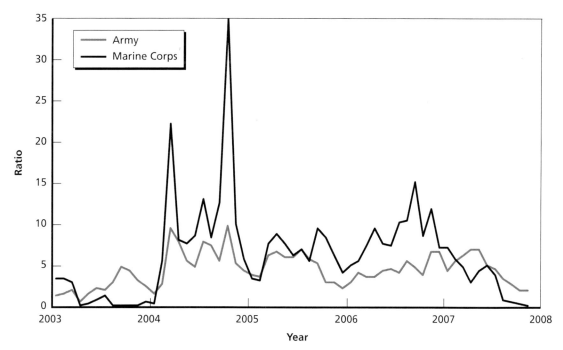

RAND *MG873-2.6*

together, but the Marine Corps ratio is higher and more variable from 2004 to 2007, reaching values of 22 in April 2004 and 35 in November 2004. The Army ratio was also highest in 2004; it declined to two in January 2006 and returned to 2004 levels by spring 2007, when the surge began. When the surge was under way, the Army ratio declined rapidly. A limitation of the ratio is that it does not account for variation in local risk. In any month, risks could have been low in some areas but quite high in others, which implies that the ratio is not an accurate measure of the risk facing a particular unit.

The rotational nature of staffing during OEF and OIF meant that a high fraction of service members deployed at some point in the three years preceding their reenlistment decision and that a service member's cumulative months of deployment would increase over time. This is seen in Figure 2.7, which traces average months of hostile deployment among soldiers, conditional on having hostile deployment. Average months doubled from 5.5 to six months in 1998 through mid-2003 to 12 months by mid-2006.

The increase in average months of hostile deployment would have been greater if the Army had not spread the burden of deployment widely across its units and troops. Figure 2.7 shows the sharp increase in the second half of 2003 in the percentage of soldiers who had been deployed in the past year, and this served to limit the increase in further deployment of those who had already been deployed. Among soldiers with any hostile deployment in the three years preceding the second-term reenlistment decision, the average number of hostile deployments increased from 1.2 in 1996 to 1.3–1.4 in 1999 through mid-2003, and to 1.5–1.6 in 2006–2007 (see Figure 2.8).[3] As shown in the figure, the pattern was similar at the first-term reenlistment point, but the average number of hostile deployments was slightly lower. These averages cover all hostile deployments, but deployments to OIF and OEF probably accounted for most of the change.

The increase in deployment intensity was not shared uniformly across services. For service members who were deployed for hostile duty in the three years prior to the reenlistment decision, Figure 2.9 shows the distribution of first-term service members in each service across four bins defined by months deployed for hostile duty: 1–6 months, 7–11 months, 12–17 months, and 18 or more months. (The proportions in the bins sum to one in each year.) For the Army, there was a clear increase in heavy deployment experience (12 or more months) starting in 2004, and in 2006, the fraction of Army personnel with 18 or more months spiked. For the Marine Corps, the increase in heavy deployment experience is less pronounced and concentrated in the 7–11 month bin. By 2006, two-thirds of those in the Army who had been deployed had 12 or more months of deployment, as had one-half of the marines. For the other branches, there was little movement in the proportion with many months deployed.

Despite the increases in average months and episodes of hostile deployment and, in the Army and Marine Corps, the proportion of service members with a high number of months of deployment, the first- and second-term reenlistment rates in all services were higher in 2002–2007 than in 1996–2001 (see Figures 2.10 and 2.11). Army and Marine Corps first-term rates held steady in 2004–2006 and increased in 2007. Air Force first-term reenlistment decreased from 57–60 percent to 53 percent but was still above pre-2002 levels.

---

[3]   We tabulated the number of service members with one, two, or three or more hostile deployments. We assigned a value of three to the three-or-more category. Only a small fraction of service members had three or more deployments, so using a value of three rather than a more precise value had little effect on average number of deployments.

Figure 2.7
Average Months of Hostile Deployment in 36 Months Preceding Reenlistment Decision, Army

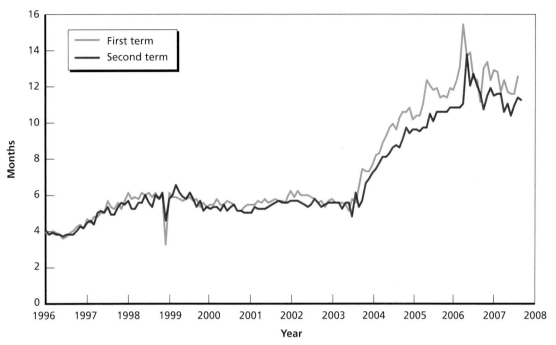

RAND *MG873-2.7*

Figure 2.8
Average Number of Hostile Deployments in 36 Months Preceding Reenlistment Decision, Army

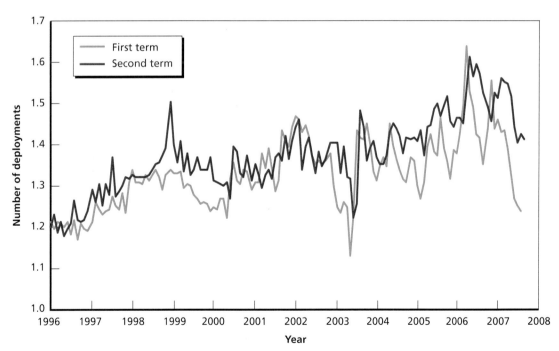

RAND *MG873-2.8*

Figure 2.9
Distribution of Months Deployed in 36 Months Preceding Reenlistment Decision, by Service, First Term

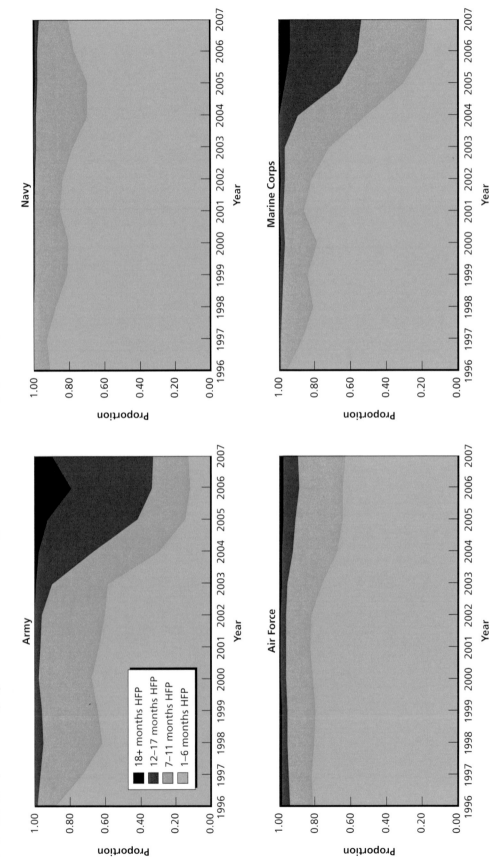

**Figure 2.10
Reenlistment Rate, by Service, First Term**

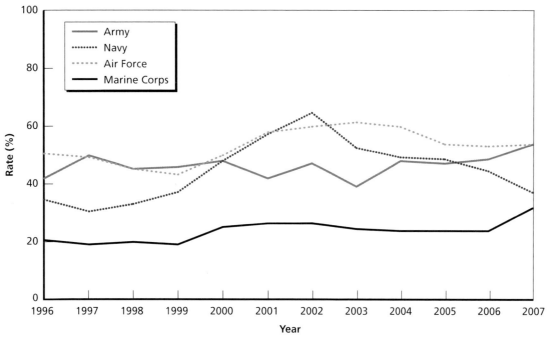

**Figure 2.11
Reenlistment Rate, by Service, Second Term**

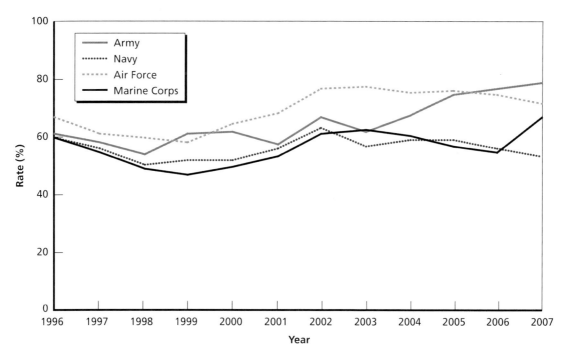

Navy reenlistment declined after 2002, but the Navy was downsizing. Its first-term rate decreased from a peak of 64 percent in 2002 to 38 percent in 2007. The second-term rate in 2002–2007 increased in the Army, was steady in the Air Force, fell and then rose in the Marine Corps, and fell from 62 percent to 52 percent in the Navy.

Literature Review

Research on 1990s deployments found that they were positively related to reenlistment. However, many first- and second-term enlisted personnel had no deployment or only one deployment (Hosek and Totten, 1998, 2002), and there were few deaths and casualties. By comparison, deployments to Iraq and Afghanistan have been marked by IEDs, suicide bombings, and attacks by insurgents and terrorists, with approximately 50,000 wounded, killed, or injured between 2002 and 2008. As discussed earlier, one deployment was common, and many service members had a second deployment; if they reenlisted, they could expect more. Mental health issues from deployment have emerged as a national policy concern, and the services and the U.S. Department of Veterans Affairs realize that they must improve care for those returning with post-traumatic stress disorder (PTSD), depression, or brain injuries and do more to prepare personnel for the mental and emotional challenges of deployment and to support their families. The studies that we review here considered the relationship between deployment and retention, how deployment may affect mental well-being, and the preparation of soldiers and families to cope with the stresses of deployment. There are few studies on deployment and retention, but the literature on mental health issues of deployment is vast, and we review only a small sample.[4]

Deployment and Retention

Studies by Quester et al. (Quester, Hattiangadi, Lee, and Shuford, 2006; Quester, Hattiangadi, and Shuford, 2006) found that reenlistment was higher for marines who deployed than for those who did not and tended to decrease as deployment days increased. The decline was more gradual in fiscal year (FY) 2005 than in FY 2004. The FY 2004 reenlistment rate for first-term marines with no days of deployment was about 33 percent. For marines with 1–100 days deployed, it was 35 percent, and it declined to 31 percent for 101–200 days deployed, 28 percent for 201–300 days deployed, 25 percent for 301–400 days deployed, 21 percent for 401–500 days deployed, and 20 percent for 500 or more days deployed. In FY 2005, the reenlistment rate for marines with no days of deployment was 26 percent, lower than the 33 percent in FY 2004. The reenlistment rate for 1–100 days was 35 percent, and it declined to 23 percent for 401–500 days deployed. The reenlistment rate for those with 500 or more days deployed was 28 percent, far above the 20 percent in FY 2004. In FY 2004, 12 percent of the marines making a reenlistment decision had no deployment days, and 10 percent had 500 or more deployment days; the percentages in FY 2005 were 8 percent and 17 percent, respectively.

---

[4]   For further reading on mental conditions affecting service members, including their prevalence, treatment, and cost, see Tanielian and Jaycox (2008), which reviews much of the literature in that area. MacDermid et al. (2008) discuss resilience in military families, and Castaneda et al. (2008) focus on the needs and support of reserve families. A survey of the literature on resilience is currently under way at RAND.

Marines with dependents had higher reenlistment rates, whether or not they had been deployed—approximately 30 percent versus 20 percent—and the reenlistment rate gap between marines with and without dependents tended to widen as deployment days increased, declining more rapidly for marines without dependents. An exception to this pattern of decline was that, in FY 2005, for marines with and without dependents, the reenlistment rate among the most extensively deployed (500 or more days) was similar to that in the least/less extensive deployment groups. More than a third of the marines in certain occupational specialties had 500 or more days deployed, including rifleman, mortarman, machine gunner, helicopter crew chief, aviation mechanic, antitank missileman, and other small specialties.

Reenlistment rates typically increased with days of deployment for second- and third-term marines in both FY 2004 and FY 2005. As in the first term, marines with dependents were more likely to reenlist than those without dependents. Among officers, retention rates were several percentage points higher for those who deployed than for those who did not, e.g., 92–95 percent versus 90 percent.

Fricker and Buttrey (2008) analyzed whether individual augmentation (IA) deployment of Navy personnel adversely affected their retention. IA deployment occurs when a service member is attached to another unit to serve with that unit during its deployment, which helps the unit to meet its manpower requirements. The study found that Navy enlisted personnel and officers with IA deployment are more likely to reenlist than those without deployment. When they constrained their IA sample to include only those deployed to Iraq or Afghanistan, the association between IA deployment and reenlistment increased. The authors caution that their IA data do not distinguish between voluntary and involuntary IA deployment, and the positive association may be, in part, a result of "volunteerism" (self-selection or targeted selection by unit commanders).

Hosek, Kavanagh, and Miller (2006) analyzed Status of Forces survey data for active-duty personnel surveyed in March 2003 and July 2003. These data contain information on the service member's stated intention to stay in the military and the level of work stress relative to the member's "usual" level of work stress, among other variables. The analysis found that the intention to stay in the military was lower if the service member frequently worked longer than the usual duty day, and this effect was statistically the same for personnel who were not away in the previous year as for those who were away. This finding is notable because of the high prevalence of working longer than the usual duty day for more than 120 days, which was the highest category in the survey question. For enlisted members away in the year preceding the survey, the percentages in this category were as follows: Army, 40 percent; Navy, 33 percent; Air Force, 29 percent; and Marine Corps, 38 percent. For those not away in the past year, the percentages were much lower: Army, 20 percent; Navy, 13 percent; Air Force, 17 percent; and Marine Corps, 20 percent. The percentages for officers were generally similar. The analysis also found that intention to stay in the military was lower if the time away was much more than expected. It was higher if the individual felt well prepared and that his or her unit was well prepared. The results for higher-than-usual work stress mirrored those for intention to stay. By and large, the same variables were significant but worked in the opposite direction. However, although variables that increased stress also reduced the intention to stay, the simple overall relationship between higher-than-usual work stress and intention to stay was positive. To explain the positive overall relationship, the authors suggested that, while a high pace of military operations increased the likelihood of higher-than-usual work stress, at the same time, it involved the service member in meaningful activities that provided some amount of satisfaction, leading to a high intention to stay. The research also involved focus groups with officers and enlisted personnel in the Army, Navy, Air Force, and Marine Corps, many of whom had returned from a deployment to Iraq or Afghanistan. The findings from the focus groups were consistent with the regression results. They also brought up the importance of communication between the deployed service member and family and friends back home, as well as the stress arising when a married member reintegrates with his or her family after deployment.

Savych (2008) studied the effect of deployment on the spouses of military personnel and included an examination of retention, which we summarize here. He used data on deployment and retention from administrative records linked to Status of Forces surveys of active-duty personnel from July 2002; March, July, and November 2003; April, August, and December 2004; and March, August, and December 2005. Using a specification that pools data across surveys and branches of service, he found that personnel deployed since September 11, 2001, are 2.3 percent more likely to reenlist, other things equal, but 2.1 percent less likely to state a positive intention to stay at the time of the survey, which may be up to two years before the reenlistment decision.[5] The negative effect of deployment on intention to stay derives from first-term personnel, who are 3.5 percent less likely to state a positive intention. Second- and higher-term personnel are only 1 percent less likely to do so, and their estimates are not statistically significant. Further, deployment effects vary by dependency status. For service members with no children or children under age 5, the effect of deployment on reenlistment is zero, while for those with children aged 6–12 or older than 12, the effect is positive. The effect of deployment varies by term of service. It is not significant for first- or second-term reenlistments, but it is positive and significant for higher terms. Finally, he found a negative relationship between spouse employment and both reenlistment and intention to stay.

Mental Health

Hoge, Auchterlonie, and Milliken (2006) used data on soldiers and marines who returned from deployment to Iraq (OIF), Afghanistan (OEF), and other locations between May 1, 2003, and April 30, 2004. Many of these service members completed the postdeployment health assessment (PDHA), a self-administered survey routinely given upon return from deployment. Its three pages of questions cover "deployment location, general health, physical symptoms, mental health concerns, and exposure concerns" (Hoge, Auchterlonie, and Milliken, 2006, p. 1024). Immediately after completing the PDHA, each service member meets with a physician, nurse practitioner, or physician assistant, who determines whether the service member should be referred for further evaluation. The PDHA includes screens for depression (two items: depressed mood and taking little interest or pleasure in doing things), PTSD (four items: reexperiencing trauma, numbing, avoidance, and hyperarousal), and other behavior suggesting increased risk of a mental health problem (e.g., interest in receiving help for a stressor, such as an emotional, alcohol, or family problem; thoughts of hurting or killing oneself; concerns about conflicts with close friends or family; or fear of loss of control or hurting another person). There are also three questions about combat experience: "Have you seen anyone killed or wounded on this deployment?" "Have you engaged in direct combat where

---

[5] Part of our analysis uses the same survey data, though we estimate models separately by branch of service.

you discharged your weapon?" and "During this deployment, did you ever feel you were in great danger of being killed?"

The authors defined an indicator of "any mental health concern" as a positive response to "little interest or pleasure; feeling down; interested in receiving help for stress, emotion distress, family problem; thoughts of hurting self; PTSD screen positive (two or more positive responses); thoughts of serious conflicts with others; thoughts of hurting someone or sense of a loss of control with others; and have sought or intend to seek care for mental health" (Hoge, Auchterlonie, and Milliken, 2006, p. 1027).

The prevalence of "any mental health concern" was considerably higher among those deployed to Iraq and Afghanistan than among those deployed to other locations: 19.1 percent, 11.3 percent, and 8.5 percent, respectively. The pattern was the same for the prevalence of any combat experiences: 65.1 percent for Iraq, 46.0 percent for Afghanistan, and 7.4 percent for other locations. Thus, mental health concerns correlated with combat experiences.

The study also found that "any mental health concern" was positively related to departure from military service within a year of returning from deployment. For Iraq, 16.4 percent of those with no mental health concern left military service, compared with 21.4 percent of those with a mental health concern. The comparable figures for Afghanistan are 12.8 percent and 20.8 percent, and, for other locations, 14.3 percent and 23.0 percent. The findings indicate that (1) leaving military service is more likely among service members with a mental health concern, but (2) among those with a mental health concern, the percentage leaving military service was slightly lower for Iraq and Afghanistan deployments than for other deployments (21.4 percent and 20.8 percent versus 23.0 percent).

We can join the prevalence and leave rates to estimate the fraction of service members who had a mental health concern and subsequently left the military. We use the following relationship: Pr(mental health concern, leave military) = Pr(mental health concern) Pr(leave military | mental health concern). We have 0.191 × 0.214 = 0.041, or 4.1 percent for Iraq, 2.4 percent for Afghanistan, and 2.0 percent for other locations. Thus, separation rates from the military among those with mental health concerns were higher if the personnel had been deployed to Iraq or Afghanistan versus other locations, and this resulted mainly from the higher prevalence of mental health concerns related to Iraq and Afghanistan deployments rather than from higher separation rates conditional on having a mental health concern.

The "Pre-Deployment Battlemind Briefing for Warriors" (Rinehart, 2008) describes what soldiers should expect to see, hear, smell, think, and feel emotionally when deployed to a high-risk combat area, such as Iraq. It describes the realities of deployments—from boredom and loneliness to attack and death—and stresses the importance of mental toughness, being a battle-buddy, listening to leaders, trusting skills and training, and staying in contact with those back home.

We have used tables on exposure to combat in Rinehart (2008) and Castro (2008) to derive Table 2.1. As the table shows, at least 92 percent of soldiers returning from a deployment to Iraq reported any exposure to combat. In comparison, Hoge, Auchterlonie, and Milliken (2006) reported that 65.1 percent of soldiers deployed to Iraq and returning in May 2003 to April 2004 had any combat experience. The higher exposure to combat in 2007–2008 and the positive association between exposure to combat and mental health concerns suggest that the fraction of soldiers returning from deployment with a mental health concern was higher in 2007–2008 than in 2003–2004.

Table 2.1
Prevalence of Exposure to Combat

Item	Percentage Reporting at Least One Occurrence During Most Recent Deployment to Iraq	
	Circa Early or Mid-2007	Circa Late 2007 or Early 2008
Knew someone seriously injured or killed	86	92
Had a member of one's own team become a casualty	75	74
Saw dead or seriously wounded Americans	65	74
Handled or uncovered human remains	50	47
Was responsible for the death of an enemy combatant	48	33
Saved the life of a soldier or civilian	21	19

SOURCES: Early and mid-2007 data are from Castro (2008, p. 2); late 2007 and early 2008 data are from Rinehart (2008, p. 22).

Adler, Castro, and McGurk (2007, p. 1) discuss psychological debriefing as an intervention technique that can be used with military personnel during deployment or immediately after deployment: "Military personnel who experience combat-related events are more likely to report mental health problems yet few early interventions have been designed to do more than assess those with problems or treat those with diagnoses." The "Battlemind debriefing" report is a component of the Walter Reed Army Institute for Research Battlemind system of training. The debriefing consists of a structured, group discussion of a particularly stressful experience, with variations in the number of phases of discussion, its focus, and the degree of structure provided to the group. The report provides guidance on debriefing procedures and offers results, based on 25 debriefings conducted as part of "the first-ever randomized controlled trial of psychological debriefing with a military population," suggesting that this intervention is "well-received by military personnel [returning from deployment to Kosovo], is not harmful, and may even be useful" (Adler, Castro, and McGurk, 2007, p. 3).

Castro (2008) observes that exposure to combat can have long-lasting effects on mental health. The self-reported prevalence of depression, anxiety, PTSD, and any of these psychological conditions was higher 12 months after deployment to Iraq than three months after deployment (see Table 2.2).[6] Furthermore, the prevalence of each of these conditions was higher after deployment to Iraq than was the case before deployment.

Deployment places strain on marriages, and self-reported marital satisfaction decreased among soldiers deployed to Iraq (Castro, 2008, p. 26). For instance, prior to deployment, 82 percent responded with "agree" or "strongly agree" to the question, "I have a good marriage." Twelve months after deployment to Iraq, this had decreased to 73 percent. Similar changes occurred in response to the questions, "My relationship with my spouse makes me happy," "My relationship with my spouse is very stable," and "I really feel like a part of a team with my spouse."

---

[6] Tanielian and Jaycox (2008) present estimates of prevalence among members of any service who were deployed to Iraq or Afghanistan: PTSD, 14 percent; major depressive disorder, 14 percent; either PTSD or major depressive disorder, 18.2 percent. That survey was conducted in fall–winter 2007.

Table 2.2
Prevalence of Mental Health Conditions Following Deployment to OIF

Condition	Pre-OIF (%)	3 Months Post-OIF (%)	12 Months Post-OIF (%)
Depression	6.3	7.9	12.0
Anxiety	6.4	7.9	11.5
PTSD	5.0	12.9	17.0
Any of the above	9.3	17.1	23.2

SOURCE: Castro (2008, p. 20).

However, a multivariate analysis of marital dissolution by Karney and Crown (2007) found that, among service members who married between 2002 and 2005, the effect of deployment on marital stability was positive, i.e., deployment was associated with a lower likelihood that a marriage would dissolve. This finding was statistically significant in both the active and reserve components. The finding suggests that self-reported information from service members about how deployment affected them may not correspond to observed behavior. Work under way at RAND is extending this analysis to a wider time frame, 1998–2008.

Family Support

Although not a study, the Army's "Predeployment Briefing" (McClintock, undated) is an example of the online resources available to military families. It includes a phone number for the Fort Carson information and referral program, which provides information on installation and community resources. It also promotes Military OneSource, a widely used portal for service members and their families for information on such topics as parenting and childcare, education, older adults, midlife and retirement, deployment and reunion, finances, legal concerns, everyday issues, work, international questions, relocation, emotional well-being, addiction and recovery, and grief and loss. The briefing also lists programs at Fort Carson addressing these topics.

The Military OneSource portal also contains links under its "Deployment" category that address military benefits, managing stress, deployment and return, single service members, and everyday military issues. Under the topics of deployment and return, there are links to dealing with deployment, preparing for deployment, reservist issues, and returning from deployment. For example, the topic of dealing with deployment includes links to articles (e.g., "Backup Care Planning for Children of Deployed Service Members," "Coping with a Deployment Extension"), booklets and recordings (e.g., "Finding Strength in Family and Community," "These Boots: A Spouse's Guide to Stepping Up and Standing Tall During Deployment"), quick tips (e.g., "Communicating During Deployment," "Keeping a Relationship Strong During Deployment"), resource guides (e.g., "Army Spouse Guide: A First Guide for Army Spouses and Family Members," "Combat Stress Resources for Military Families"), Web links (e.g., "Afterdeployment," a mental wellness resource with links to brief self-assessment tools), and worksheets and checklists (e.g., "Discussing Danger While a Parent Is Away," "Helping Children Cope During Deployment").

CHAPTER THREE

Modeling Deployment and Reenlistment

Reenlistment, an occupational choice decision, can be modeled as a comparison of the utility of staying in the current occupation with the utility of the best alternative. If the utility of serving in the military is higher than that of working in a civilian job, the individual will choose the military over the civilian job. This choice involves a comparison of expected utility based on information available at the time of the decision. Holding civilian utility constant, reenlistment is more likely to occur if the military utility is higher.

In applying this idea, we assume that utility depends on the quantity of deployment, the qualitative difference between deployment time and home time, and deployment-related pay. In our model, a service member has a preferred level of deployment, and when expected deployment differs substantially from preferred deployment, the service member is more likely not to reenlist. However, added pay can compensate for suboptimal deployment, whether it is too much or too little. We use the model to motivate our empirical work in Chapters Six and Seven and to consider the role of bonuses as an offset to the decreasing effect of deployment on reenlistment in Chapter Eight.

Utility Model of Deployment

We assume that a service member's utility depends on time at home station, time deployed, and income. Duty hours at home station and during deployment are set by the service. Home time becomes zero when deployment occurs, which reduces utility, but deployment may increase utility, and deployment pay, if offered, will increase utility. The inclusion of deployment in the utility function is consistent with the finding that nonhostile deployment has a direct, positive effect on reenlistment, even when there is no deployment-related pay.[1] We hold the share of time on nonhostile deployment constant and focus on hostile deployment. If none of the remaining time were allocated to hostile deployment and all were allocated to home time, the marginal utility of home time would be relatively low, assuming diminishing marginal utility of home time. Also, for members preferring some hostile deployment rather than none, the marginal utility of hostile deployment time would be relatively high. Utility is maximized where the

---

[1] In particular, nonhostile deployment includes no deployment-related pay for members without dependents, with the exception of sea pay, which is paid to sailors and marines on sea duty. We ran separate models (not shown) for members with and without dependents, by branch of service. The effect of nonhostile deployment on reenlistment was positive and significant in all cases.

marginal utilities of home time and hostile deployment time are equal.[2] If hostile deployment time were increased from this point, the increase in utility from hostile deployment would be less than the decrease from home time. If hostile deployment time were decreased, the decrease in utility from hostile deployment would be greater than the increase from home time.

The following example illustrates the model.[3] We define the utility function as follows:

$$U = k - \frac{\left(\left(1 - d_n - d_h\right)^\alpha + \theta d_n^{\beta_n} + \phi d_h^{\beta_h}\right)^{1-\delta}}{1-\delta} \frac{\left(\left(1 - d_n - d_h\right)y + w_n y d_n + w_h y d_h + b y\right)^{1-\gamma}}{1-\gamma}.$$

The term $\left(1 - d_n - d_h\right)^\alpha + \theta d_n^{\beta_n} + \phi d_h^{\beta_h}$ is effective time, where d_n and d_h are the fractions of time in nonhostile and hostile deployment, and the parameters θ and ϕ adjust for the possibility that time during nonhostile or hostile deployment is qualitatively different from home time. Any deployment involves separation from family and friends, and hostile deployment also involves arduous duty and danger. The term for military income, $\left(1 - d_n - d_h\right)y + w_n y d_n + w_h y d_h + b y$, depends on the mix of home and deployment times; military compensation when not deployed, y; military compensation for nonhostile deployment, $w_n y$; military compensation for hostile deployment, $w_h y$; and bonus, by. Deployment pay rates are written relative to y, the value of which is normalized to 1. The pay rate for hostile deployment is greater than the pay rate for nonhostile deployment, so $w_h > w_n > 1$. The utility function has constant relative risk aversion, as commonly assumed in consumer models.[4]

For selected parameter values, utility has an inverted U-shaped relationship to hostile deployment time, as shown in Figure 3.1.[5] Here, utility is maximized when the fraction of time on hostile deployment is 0.22. If the service member could choose, he or she would prefer to be deployed for hostile duty 22 percent of the time. Utility is not much lower than the maximum when hostile deployment is between 10 and 40 percent. However, utility decreases rapidly as hostile deployment decreases from 10 to 0 percent or increases beyond 40 percent.

The service member is not free to choose deployment time but is assigned to a unit and deploys when the unit deploys.[6] Deployment patterns differ by branch of service and, within a service, by occupational specialty. At enlistment, an individual can choose the branch of

---

[2] There are also corner solutions. If the marginal utility of home time is always greater than the marginal utility of hostile deployment, the individual prefers to never deploy. If the marginal utility of hostile deployment is always greater than that of home time, then the individual prefers to be deployed always. The discussion can be expanded to allow nonhostile deployment time to vary. Utility is maximized where all three marginal utilities of time are equal. Allowing nonhostile time to vary adds little insight to the discussion.

[3] The example assumes that the utility function is multiplicative in time and income, but the illustrated results do not depend on this assumption. Similar results can be obtained from a utility function that is additive in home time, nonhostile deployment time, hostile deployment time, and income, for instance.

[4] The measure of constant relative risk aversion is $-mU_{mm}/U_m$, where m is income. Here, $-mU_{mm}/U_m = -yU_{yy}/U_y = \gamma$.

[5] The parameter values are $k = 0$, $\alpha = 1$, $\beta_n = 0.5$, $\beta_h = 0.3$, $\gamma = 1.1$, $\delta = 1.2$, $\theta = 0.9$, $\phi = 0.8$, $d_n = 0.1$, $y = 1$, $w_n = 1.1$, $w_h = 1.45$, and $b = 0$. The value of the risk aversion parameter γ accords with empirical estimates (see, e.g., Hurd, 1989).

[6] This is a generalization. A possible exception would be an active-duty service member who wants to deploy (or redeploy) and requests transfer to a unit that is scheduled to deploy. Similarly, a reservist has some limited control over deployment by withdrawing from a unit well in advance of when it is scheduled to deploy or by requesting assignment to a unit that is deploying.

Figure 3.1
Utility and Hostile Deployment Time

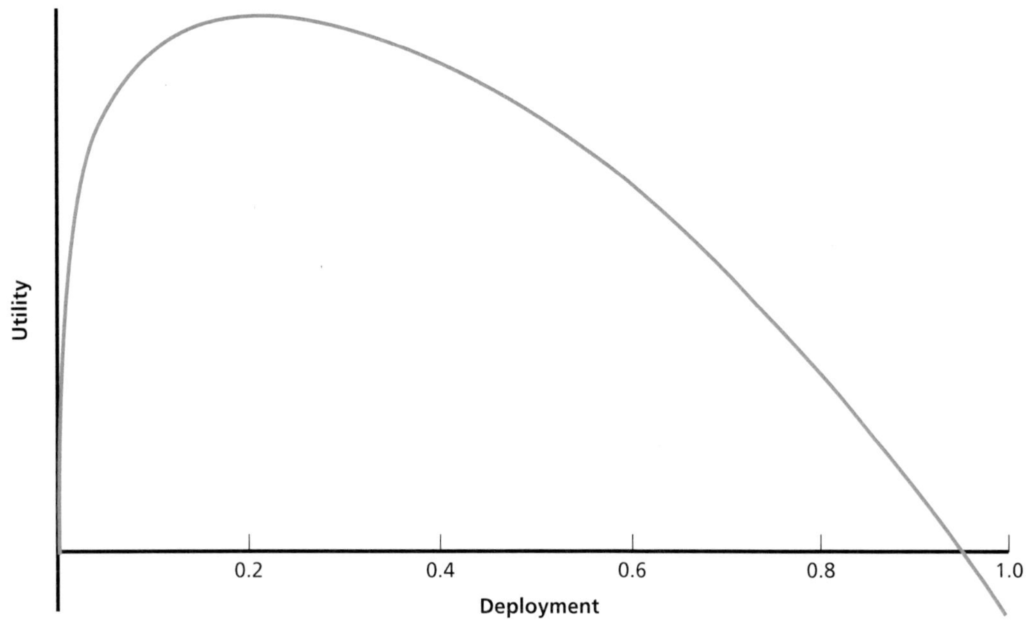

Utility

0.2   0.4   0.6   0.8   1.0

Deployment

RAND *MG873-3.1*

service and has some choice in occupational area and specialty, but once enlisted, the individual's deployment is governed by the service and not chosen by the individual.[7] As a result, realized deployment and expected future deployment may differ from both the individual's preferred level of deployment and the initially expected level of deployment. Although expected future deployment may not equal the preferred level, it must nevertheless correspond to an expected utility of military service that is higher than the expected utility of the best alternative; otherwise, the individual would not be willing to reenlist.

This discussion suggests four implications about the effect of deployment on reenlistment. First, expected deployment that is not too far from the preferred level may have little effect on utility and therefore little effect on reenlistment. Second, assuming that the past realization of deployment affects expectations, if service members typically prefer some deployment, then reenlistment will be higher for service members who had some but not too much deployment than for service members who were not deployed. Third, reenlistment will be lower for service members with extensive deployment relative to expected deployment. Fourth, reenlistment will be lower the more uncertain deployment is relative to the expected level. The latter point follows from the convexity of the relationship between utility and deployment time, which implies that the individual is risk-averse to uncertainty in deployment time.

In relating the model to our empirical work, we recognize that preferences for deployment may vary by individual. This implies that members will sort into service branches and occupations, in part, on the basis of their preferences for dangerous deployments. For example, individuals desiring to take part in ground operations would join the Army or Marine Corps rather

---

[7] An individual who wants to deploy may try to arrange a transfer to a deploying unit. The transfer is unlikely unless the deploying unit has an opening for the individual's specialty and grade and the sending unit is willing to give up a member.

than the Navy or Air Force and would enter a combat-arms specialty rather than personnel or administration. Consequently, members whom we observe with heavy deployment experiences might be those who have greater tolerance for the hardships associated with deployment. Alternatively, the individuals who seek out occupations with a high level of deployment exposure might do so because of the financial incentives (such as enlistment bonuses) that encourage entry into these occupations. Either of these mechanisms suggests that deployment may be related to preferences for reenlistment. A related point is that the effects of deployment that we estimate will be weighted toward the effects for individuals who tend to sort into high-deployment occupations or branches.

Although we do not explicitly model preference heterogeneity, we do report results stratified by dimensions that are highly related to ex ante expectations of deployment risk. In particular, we stratify all estimates by branch of service and first or second term. We also report some results stratified by other characteristics associated with preferences, such as the combat-arms specialty, marital status, and gender. By comparing results across these groups, we obtain indirect evidence of how the effects of deployment vary by preferences for deployment, though it is worth reiterating that we do not claim to directly observe preferences.

Further, if deployment is assigned to the individual by the service based on its occupational needs, it should be largely independent of the individual's preference for deployment, given occupation and service. This is consistent with the idea, which seems accurate, that the service chooses the unit to deploy without considering the preferences of individuals in the unit.[8] At the end of a term of service, differences in the expected utility of future deployment will likely depend on (1) the deployment exogenously assigned to the individual (a natural experiment) and (2) the individual's learning about the utility or disutility of deployment. In empirical work, the use of fixed effects for occupational specialties accounts for preference-related selection into the occupation at enlistment. Further, if selection changes over time, year effects and, better still, occupational specialty-quarter fixed effects should help control for this.[9]

---

[8] Lyle (2006) argues that unit deployment is based on division- and brigade-level decisions. Lyle (2006) and Savych (2008) use unit deployment as an instrument for individual deployment and obtain deployment effects that are quite close to those based on individual deployment. Both Lyle and Savych use a Hausman test and reject the hypothesis that individual-level deployment is not orthogonal to the error.

[9] Although we have cast our model as an expected-utility model that is updated with experience, it could be recast in terms of the theory of planned behavior in social psychology (Ajzen, 1985, 1991). In this formulation, the individual has behavioral beliefs about the consequences of reenlisting, including the occurrence of deployment and its related effects. The beliefs depend on the individual's perception of how friends, family, and other "influencers" judge this behavior (normative beliefs) and perception of social norms toward the behavior (subjective norm). Given the set of beliefs, the individual forms an attitude toward the behavior, i.e., a positive or negative evaluation of reenlisting. Another element of the theory, control over being able to perform the behavior, is not an issue here; the individual is given the opportunity to reenlist. Based on beliefs and attitude, the individual forms a behavioral intention, and it is a precursor of actual behavior. The model allows for updating, and it recognizes that, in empirical work, intentions might not be followed by behavior if unobserved factors affecting beliefs and attitudes change after intentions are reported.

CHAPTER FOUR

Data Sources and Analysis Samples

We employ two datasets in our analyses. The first is a file generated from U.S. Department of Defense (DoD) administrative files and comprises all reenlistment decisions made by enlisted service members between 1996 and September 2007. This dataset contains information on deployment prior to the reenlistment decision, a measure of the reenlistment bonus for which the service member would be eligible, details about his or her status in the military (e.g., service, pay grade, years of service), and demographic characteristics. The large number of observations in the dataset makes it possible to conduct analyses focusing on subgroups of particular interest and on narrow periods of time.

The second dataset consists of Status of Forces surveys linked to the above-mentioned file. The survey data include subjective measures of a respondent's well-being in the military that are not available in the administrative data, including intention to reenlist, whether the individual spent more or less time away from home than expected, information on work intensity, and whether the individual had unusually high levels of work stress or personal stress. By linking the survey data to the administrative data, we could include more complete information about deployment and, importantly, actual reenlistment outcomes.[1]

Below, we describe the source data and discuss how we generated the decision-level data file. We also describe the survey-administrative linked data.

Data Sources

Proxy PERSTEMPO File

The Defense Manpower Data Center's (DMDC's) Proxy PERSTEMPO (personnel tempo) file was our main data source on deployments, reenlistment decisions, and service-member characteristics. This file consists of individual-level records on active-duty service members. It has longitudinal information on military occupational specialty (MOS), family status, education, pay grade, measures of deployment (described in greater detail later), and the time remaining on the current term of service. The PERSTEMPO file also specifies gender, date of birth, race, ethnicity, and Armed Forces Qualification Test (AFQT) category.

---

[1] The Status of Forces survey results contain information on deployments, but the exact questions varied across survey waves. The administrative data on deployment measures, in contrast, were constructed in a consistent manner across all survey respondents.

Joint Uniform Military Pay System (JUMPS) File

The JUMPS file contains disaggregated information on the types of pay that service members receive each month, including whether a member received a reenlistment bonus. The records indicate the size of the total reenlistment bonus as well as the amount paid out in a particular month. We use the total amount of the bonus and the member's pay grade to infer the bonus multiple (or "step"), which indicates the amount of the bonus offered.

Status of Forces Surveys and Linked Survey-Administrative Record Dataset

DMDC conducts the Status of Forces survey of military personnel to study trends in personnel well-being and to collect information on topics that are of particular interest to policymakers. In this analysis, we used the 10 waves of the Status of Forces Survey of Active Duty Personnel, which were administered between July 2002 and December 2005. Survey respondents are selected via stratified random sampling, and the survey questionnaire is administered over the Internet. Around 35,000 individuals (officers and enlisted personnel) are sampled for each survey wave, and among these, about 10,000 complete the survey. Nonresponse is more common among junior and enlisted personnel than among senior personnel and officers.[2] DMDC produces weights that adjust for the sampling design as well as for survey nonresponse. These weights are used in tabulations of the survey data reported here. However, it should be noted that the survey weights can account only for differential nonresponse based on observable covariates; nonresponse driven by unobservable factors may still lead to bias. When estimating regression models, survey weights are not used because the results were stratified by some of the key variables used to generate the weights (such as branch of service and pay grade) or were included as regressors (such as gender and race). Moreover, we obtained similar estimates when we applied the survey weights (results available upon request).

DMDC provided RAND with specialized survey files that included the encrypted social security numbers and selected survey items.[3] Information from the administrative data was then merged with the survey file on the basis of the encrypted social security number.

Dataset of Reenlistment Decisions

The main analysis file that we used to study the relationship between deployment and retention consists of all reenlistment decisions made between 1996 and September 2007. Whether an individual reenlists was then modeled as a function of deployment experiences, bonuses, and other characteristics. In this section, we describe how we generated this dataset, starting with how reenlistment decisions are identified and classified in the data.

Identifying Reenlistment Decisions

Because we did not have reenlistment contract records, reenlistment decisions were inferred from monthly information about the time remaining to expiration of term of service

---

[2]   We exclude cases that had missing data on key constructs such as intention to reenlist. The results may have been somewhat different had we imputed values for missing data, though we do not know in which direction the results would have changed.

[3]   Due to confidentiality concerns, only the variables that we determined to be necessary for this project were included in the files provided to us by DMDC.

(ETS). This approach was first used in Hosek and Totten (2002) in a study that also used PERSTEMPO data to study the effects of deployment and time to promotion on reenlistment in the period from 1996 to 1999. The algorithm used to identify the timing of a decision and what the decision was is briefly described here; see Hosek and Totten (2002) for a more thorough discussion.

To identify decisions, individuals are followed over time in the PERSTEMPO data. As the ETS date approaches, members make one of three decisions. They can reenlist, exit the military, or extend their contract. Extensions and reenlistment differ in a qualitative sense. Extension is a way to postpone the reenlistment decision if the individual is uncertain. Past research has shown that reenlistment and extension decisions are sensitive to macroeconomic conditions (Hosek and Peterson, 1985). In addition, extensions might be used strategically to time a reenlistment so that it occurs while the service member is deployed, at which point any reenlistment bonus would be subject to preferential tax treatment (if the deployment is to a combat-zone tax-exclusion, or CZTE, area). It is also a way of prolonging time in the service until a convenient time of departure, e.g., at the end of a school year.

If a service member exits, the months remaining to the ETS date will fall to zero in the PERSTEMPO data and the pay-grade variable will be missing, indicating that he or she left the military and no longer receives income from the military. Reenlistments and extensions, in contrast, will both result in an increase in the months to ETS. The PERSTEMPO data do not identify whether a member extends or reenlists. We define an extension as an increase in months to ETS of 23 months or fewer and a reenlistment as an increase of 24 or more months. Here, we focus on whether someone exits the military or reenlists. Therefore, if a member extends, he or she is not considered to have made a reenlistment decision and is instead followed until exit or reenlistment.[4]

Creation of the Analysis File

Having defined reenlistments in this fashion, we searched through the PERSTEMPO file for instances of reenlistments or exits and constructed an analytic file consisting of records of these decisions. Individuals who reenlist at the end of their first term may show up once again in the file if their second-term reenlistment decision is observed before the end of the data (September 2007). Because individuals who reenlist twice generally become career military personnel, we do not include decisions made at the third or higher decision point. However, similar results were obtained when all decisions beyond the first one were grouped together.

In addition, we made three other sample restrictions. First, individuals with missing data for AFQT category, race, or education were dropped. Second, members who exited the military more than six months from the end of their ETS were excluded. Most attrition occurs within the first year of service; exiting early is often a sign of an involuntary separation or a separation due to unusual or extenuating circumstances that are unlikely to be driven by deployment

---

[4] The 24-month cutoff is somewhat arbitrary, and reenlistment bonuses are not offered to individuals reenlisting for less than 36 months. However, we use a 24-month cutoff in the interest of continuity with earlier research (Hosek and Totten, 1998, 2002). In addition, the results are not especially sensitive to this definition. If we exclude the approximately 30 percent of the sample of first-term reenlisters whose ETS increases by less than 36 months, the estimated deployment effects are similar to those with the full sample.

experiences.[5] Finally, individuals are excluded if they had completed fewer than three years of service at the time of their reenlistment decision. This restriction was made because members with fewer than three years at the time of their decision (and who did not exit early) typically signed an initial service contract for two years. Some of our analyses examine the effect of deployments that occur during the three-year period prior to the decision, and individuals who have fewer than three years of completed service at the time of their decision will therefore have an artificially low number of deployments or months deployed.

Deployment Measures

The PERSTEMPO file has several measures indicating monthly deployment status.[6] One is a flag for whether an individual was away on deployment in a particular month. A second flag indicates whether an individual received HFP in a given month. HFP is paid when a member is stationed in a location deemed to be hostile. Receipt of HFP implies that the service member was deployed, but individuals deployed to a nonhostile setting do not receive HFP. Thus, service members were classified into one of three mutually exclusive states in any given month: deployed to a hostile location, deployed to a nonhostile location only, or not deployed.

The information on deployments is unfortunately crude in that we know little about the circumstances of a particular deployment. For instance, members stationed in a support role in Kuwait or serving on the front lines in Iraq would all receive HFP and be treated the same way in the analysis, even though the nature of their deployments and the risks that they faced may have been radically different. Nonetheless, the availability of monthly data on deployments allows us to generate detailed measures of the number of deployments and the number of months deployed. The number of months deployed during a certain period was computed directly from the underlying data. The number of deployments was calculated as the number of uninterrupted spells deployed during a particular period, including spells that began before the start of the period and those that were still ongoing at the end.

The deployment variables used in the analysis are based on two different time intervals, one equal to 12 months and the other equal to 36 months, each ending three months prior to the decision. The three-month buffer between the end of the observation window and the decision is used to guard against bias due to reverse causality. Specifically, preferential tax treatment of reenlistment bonuses implies that members have incentives to time a reenlistment so that it coincides with a deployment by extending their contract so as to reenlist while deployed. With no buffer, the service member would have an additional deployment added to his or her

---

[5] If service members exited early to avoid deployment, then we would expect to see an increase in early exits starting in 2002. To investigate this possibility, we examined DMDC data on enlisted continuation rates by service for year 1 to year 2 and for year 2 to year 3. There is no decrease in continuation in any of the services, with the exception of the Air Force in 2004 and 2005, where the continuation rate decreased from a high of 93 percent in 2003 to a low of 89 percent in 2005. The other services had fairly stable continuation rates, as did the Air Force in other years.

[6] Another source of information on deployments is available from DMDC's Global War on Terrorism Contingency File. This dataset contains records for each deployment and activation made in support of the global war on terrorism (GWOT). We did not use these data for our analysis because they begin with deployments after September 11, 2001, whereas we are interested in examining how the relationship between deployment and reenlistment changed before and after 9/11. The data also include only data on deployments in support of the GWOT. Notably, our contingency file extract does not have usable information on the exact location where a deployment took place. In particular, it does not show whether a deployment was to Iraq or Afghanistan.

count of deployment spells, which would induce an upward bias in the estimated effect of deployment on retention.

In the analysis, our initial specification for deployment is an indicator variable for being deployed at some point in the prior year. To examine whether there are differential effects in the length of time on deployment or the number of deployments, we also estimate models allowing for the deployment effect to vary by the total number of months on deployment or on the number of deployment spells a member experienced. For these analyses, we use the 36-month window because a 12-month window is too short to capture multiple deployments.

Reenlistment Bonuses

Selective reenlistment bonuses (SRBs) are intended to help the services maintain the desired force level in particular occupational specialties. SRBs are set at the MOS-by-zone level: Zone A is 21 months to six years of service, zone B is 7–10 years of service, and zone C is 11–14 years of service. The formula for the size of an SRB is the product of term length (the years of additional service obligated by the new contract), monthly basic pay at the time of reenlistment, and the bonus multiplier, or step, which is equal to zero when no SRB is offered.

We use the multiplier as a measure of reenlistment bonus generosity. The DMDC does not have a file on bonus multipliers, so we inferred the multiplier from the observed bonuses paid to members reenlisting in our study period. First, we determined whether each individual who reenlisted received an SRB and if so, the size of it. Then we calculated the multiplier based on the previously discussed SRB formula. Next, we calculated the average multiplier in cells determined by zone, MOS, and quarter of year.[7] This average was the estimate of the bonus multiplier that an individual would face when he or she made the decision to reenlist. Note that, by definition, the bonus measure that we use in the analysis does not vary at the individual level within the cells used to identify bonuses.

There is likely to be some slippage in the bonus measure for at least two reasons. First, the bonus multiplier could change within a quarter, whereas our calculations assume that all reenlistments within a quarter are eligible for the same SRB. Second, there may be some error in the calculation of the additional service obligation arising from a discrepancy between the month in which a service member actually reenlists (for the purposes of determining the SRB) and the month identified in the data as the reenlistment month. Although we expect both types of error to be small, the existence of such measurement errors could lead to some attenuation bias for the estimated bonus effects. But as we describe in the following chapter, there are other, more serious threats to the validity of the estimated effect of reenlistment bonuses.

Covariates and Key Subgroups

The PERSTEMPO file has extensive information on personnel, which we use as covariates in the regression analyses that follow. These include years of service, ethnicity, AFQT category, and education level. To control for whether a service member has been promoted quickly, we create an indicator for speedy promotion equal to one if his or her pay grade is higher than the average pay grade of individuals in the same branch of service who have the same number of years of service. Because the effects of deployment are likely to vary depending on a service

---

[7]   We used the three-digit DoD duty occupational codes for these calculations.

member's characteristics, we also use the PERSTEMPO file to identify key subgroups, which we use to stratify the estimates. For instance, we produce estimates by gender and also by marital status. Another important subgroup is the combat-arms occupational specialty. As noted earlier, we do not have specific information on the conditions faced during deployment, but individuals in combat likely have the most dangerous deployments.

Survey-Administrative Linked Dataset

The survey dataset pools the 10 waves of Status of Forces surveys conducted between 2002 and 2005. Restricting the sample to completed survey responses of enlisted personnel provides about 63,000 total usable observations. As described earlier, information from the personnel records (such as the deployment measures) is merged from the PERSTEMPO file. In addition, the surveys collect information that we use as additional regressors, including whether the service member spent more or less time away than expected, the number of days spent working overtime (grouped into categories), whether the service member felt well prepared, and whether the service member felt that his or her unit was well prepared for its mission.

The survey data are used to analyze four outcomes. The first two are whether the member experienced more work stress or more personal stress than usual in the 12 months preceding the survey.[8] Heightened stress is one mechanism through which deployment could affect reenlistment (Hosek, Kavanagh, and Miller, 2006). The third outcome is whether the respondent indicated an intention to stay in the military if eligible to do so. The five categories of response ranged from very unlikely to very likely to stay, and we coded "likely" or "very likely" as indicative of an intention to stay. The final outcome is whether a service member actually reenlists, and it is derived from the PERSTEMPO file. We do not observe reenlistment decisions for individuals whose ETS date is after the final month available in the PERSTEMPO file.

---

[8]   Self-reported stress is measured on a five-point scale. Respondents were asked whether they had experienced much less, less, about the same, more, or much more stress than usual.

CHAPTER FIVE

Econometric Model

This chapter presents the econometric model used in this study. It also discusses the nature of possible biases in estimates of deployment and bonus effects and our approach to mitigating them. This discussion focuses on empirical reenlistment models, but the same issues apply to the outcomes available in the survey data.

The basic model relates reenlistment of member i, R_i, to deployment, D_i (which may be a vector of deployment variables); the SRB multiplier, B_i; and a vector of covariates, X_i:

$$R_i = \theta D_i + \beta B_i + \gamma X_i + \varepsilon_i. \tag{5.1}$$

Unobservable factors that determine reenlistment are captured by the residual ε_i. If the deployment and bonus variables are uncorrelated with the residual, then standard multivariate regression produces consistent estimates of the model's parameters. In the analysis, we estimate this model using least squares regression with a binary dependent variable (i.e., a linear probability model).

However, neither deployments nor bonuses are randomly assigned. Deployment levels vary across occupational specialty, as discussed in Chapter Two.[1] Since reenlistment rates likely would vary across occupational specialty regardless of differing levels of deployment, our primary econometric specification includes controls for MOS. The large samples available in the administrative data allow us to control for a complete set of three-digit MOS fixed effects. In the analysis of the survey data, we control for one-digit MOS fixed effects. (Results using a specification that controls for three-digit occupation effects are available upon request.)

The inclusion of MOS fixed effects in the models is not sufficient for consistent estimates of the effect of deployment if selection into deployment within an occupation is related to factors that determine reenlistment decisions. However, studies of the effect of deployment on the academic outcomes of children (Lyle, 2006) and on spousal income (Savych, 2008) have shown that deployments within an occupation seem to occur as if they were randomly assigned.[2] In these studies, a Hausman test does not reject the hypothesis that deployment is orthogonal to the error term (Lyle, 2006, p. 332; Savych, 2008, p. 52).

---

[1] Deployment rates also vary by service branch and whether a service member is in the first or second term of service. However, all of our analyses are conducted at the branch-by-term level.

[2] These studies found similar results when using unit deployment as an instrumental variable for individual deployment. Since the determination of which units are deployed is based on the needs of the service, these findings support the view that deployment is not systematically related to unobserved factors at the individual level, e.g., the reenlistment preference of the individual.

The way in which reenlistment bonuses are allocated, on the other hand, does pose a serious methodological challenge. All else equal, bonuses will be higher in occupations in which policymakers anticipate having difficulty meeting retention targets. For example, in occupations for which job opportunities outside the military have improved, bonuses may need to be increased to maintain the desired staffing level. The endogeneity of the SRB implies that the simple association between the bonus level and the reenlistment rate is unlikely to be informative about the effect of the bonus on reenlistment. In fact, this association could even be negative, which, taken literally, suggests that higher bonuses lead to lower retention rates.

As with that for deployments, our approach to mitigating the bias from the endogeneity of bonuses is to control for occupation fixed effects. This approach accounts for permanent differences across occupations in reenlistment propensities.[3] But unlike deployments, the level of the SRB does not vary across individuals within an occupation at any given point in time.[4] Therefore, to identify the bonus effect, the occupation fixed-effects estimates rely on variation in the bonus level over time within an occupation.

There are two reasons that the bonus setter would change the bonus in an occupation. The first is a change in *demand* for personnel in that occupation. If demand increases, the bonus setter increases the bonus to increase retention. The second is a change in the *supply* of personnel willing to reenlist. For example, holding demand constant, if there were a surge in civilian-sector demand for workers with the skills developed in a particular MOS, the bonus setter would need to increase the bonus. Appendix A describes models of bonus setting and discusses the possible biases on estimates of the bonus effect.

Because changes in bonus resulting from changes in occupation-level demand are unrelated to the underlying willingness of service members to reenlist in the occupation, demand-induced changes in bonuses could be used to cleanly identify the effect of the reenlistment bonus. In contrast, changes in bonuses driven by supply shocks are subject to the same endogeneity concerns described earlier for cross-sectional variation in bonuses.

Thus, the credibility of the estimated effects of bonuses on reenlistment rest on whether the within-occupation changes in bonuses are largely in response to demand or supply shocks. Although we do not have any way of determining which factor is more important, there is no a priori basis for believing that supply shocks are not an important factor in producing changes in within-occupation bonus levels. Therefore, while controlling for occupation fixed effects mitigates the downward endogeneity bias present in simple cross-sectional estimates, there might still be a bias in the fixed-effects estimates. This point is formalized in Appendix A.[5]

---

[3]   This point is known from previous work; see the survey by Goldberg (2001). Hattiangadi et al. (2004) also use within-occupation variation in reenlistment bonuses to identify the effect of bonuses on reenlistment in the Marine Corps. Hansen and Wenger (2002, 2005) use occupation groups but also report results for specifications with fixed effects for occupational specialties (ratings) in the Navy.

[4]   Bonuses can vary within an occupation across service zones. However, service members are almost all in zone A when making the first-term reenlistment decision and are in zone B when making the second-term reenlistment decision. Because we produce separate estimates for first- and second-term service members, there will be very little if any within-occupation variation in bonuses at the individual level at a given point in time.

[5]   Specifically, in a model in which the bonus setter has a constant reenlistment target and occupation-level reenlistment shocks are serially correlated, (1) bonuses will be higher in occupations in which reenlistment would be below the target rate in the absence of the bonus, and (2) bonuses will be adjusted upward in response to negative reenlistment shocks. Both sources of endogeneity would lead to a downward bias in the estimated reenlistment effect. Controlling for occupation fixed effects eliminates the first source of endogeneity, reducing the severity of the overall bias. However, the second source of endogeneity remains.

An alternative approach to estimating the bonus effects is to isolate an exogenous source of variation in bonuses. This method, known as instrumental variables, requires a variable (the "instrumental variable") that is correlated with bonuses but otherwise unrelated to reenlistment. Finding such an instrument is not a straightforward process. The preceding discussion suggests that shocks to the supply of potential reenlistees might be a suitable instrument. One candidate instrument that we explored was the size of the cohort coming up for reenlistment in a given year. If an occupation had more individuals than usual coming up for reenlistment in a particular year, bonuses could be decreased without any shortfall in retention. Bonuses would be increased whenever an occupation had an unusually small number of members coming up for reenlistment. Unfortunately, we had to abandon this approach because cohort size had only a weak relationship with reenlistment bonuses.[6]

Such bias could also spill over to the estimates of the coefficient on deployment. This would be true if bonuses are correlated with deployment conditional on the other covariates included in the estimation. To see this, suppose that deployments and bonuses are positively related. Without controlling for bonuses, the positive effect of bonuses on reenlistment would be inappropriately attributed to deployment, resulting in an upward bias in the estimated deployment effect. Including the bonus as a covariate in the regression would solve this problem, provided that the effect of the bonus was itself correctly estimated. If, on the other hand, the bonus effect were underestimated, some of the positive effect of the bonus on reenlistment would load onto the coefficient on deployment, again resulting in upward bias. The mathematical justification for this intuition is presented in Appendix B.

The extent to which the bias in the estimates of the bonus effect also contaminates the estimates of deployment effect depends on the correlation between deployment and the bonus net of other covariates. However, the empirical results suggest that any bias that spills over from biased estimates of the bonus effect is likely to be small. We regressed the bonus multiplier on the fraction of individuals who were deployed in the prior year, on year dummies, and on occupation fixed effects.[7] When the data are pooled across services, there is a positive and statistically significant relationship between deployments and reenlistment bonuses. Examining this pattern by branch of service, on the other hand, reveals that this association is limited to the Marine Corps. In the other three branches, the relationship is either negative or small in magnitude and statistically insignificant.

In light of these results, we do not view the potential bias of the reenlistment bonus effect (in models that control for occupation fixed effects) as posing a serious threat to the validity of the effect of deployment. Nonetheless, to guard against this possibility, we also produce estimates from a specification that controls for MOS-by-quarter fixed effects. By focusing on comparisons between members from the same occupation making their reenlistment decision at approximately the same point in time, all of the time-series variation in the bonus is effectively purged. This has the benefit of ensuring that the deployment effects are not contaminated by biased estimates of the bonus effect. The obvious drawback of this specification, however, is that it does not produce any estimates of the bonus effect.

---

[6] An additional concern is that cohort size might be related to reenlistment beyond any influence it has on reenlistment bonuses. For example, occupations that have unusually high early attrition from the military would have smaller cohorts coming up for reenlistment, but the high attrition might be indicative of a poor retention environment.

[7] Results are available on request.

Finally, although concern about bias in the bonus effect has been driven by bonus-setting behavior in response to anticipated shortfalls or excess in reenlistment, another source of bias may be present. If a service is increasing its size selectively in certain occupations, or downsizing selectively, then it could be that (1) effort by career counselors is reallocated away from occupations that are not growing and toward occupations that are growing; (2) additional incentives, such as choice of location or preference for further training, may be reallocated in the same way; or (3) the number of positions open in occupations that are shrinking is decreased, creating a demand constraint. Such changes would be correlated with bonus but are unobserved in our data. This would result in an omitted variable bias that caused the estimated bonus effect to be biased upward. This is in contrast to the "traditional" downward bias from bonus-setting behavior. In our data period, the Navy was downsizing after 2001, and the Marine Corps and Army were growing after 2004, so upward bias may be a concern for them—along with the possibility of downward bias.

CHAPTER SIX

Empirical Results Using Survey Data

This chapter presents findings based on the survey data. We examine measures of subjective satisfaction and quality of life, as measured by higher-than-usual work or personal stress, and intentions to reenlist. Then, we analyze actual reenlistment behavior among the subsample of survey respondents for whom we observe reenlistment decisions.

Baseline Estimates

Tables 6.1 and 6.2 show the "baseline" estimates of the deployment effects for first- and second-term-plus respondents, respectively. The estimated coefficient is on an indicator for having hostile deployment in the year prior to the survey.[1,2] As discussed in Chapter Three, the Status of Forces active-duty surveys used in the analysis were administered in 2002–2005 and linked to administrative data on personnel through 2007. The results for first-term respondents provide some evidence that deployments place a heavy burden on service members. In all branches except the Air Force, deployments are associated with a statistically significant increase in the likelihood that the respondent reported experiencing greater-than-usual work stress. Greater-than-usual personal stress was significantly related to being deployed in the Army and in the Navy. These effects are largest for the Army, with an increase of 12 percentage points in the likelihood of higher-than-usual work stress and 7 percentage points in the likelihood of higher-than-usual personal stress.

The results for intention to reenlist also indicate that hostile deployment adversely affects the respondent's desire to remain in the military. For all four branches, hostile deployment has a negative and statistically significant relationship with the likelihood of intention to reenlist. This effect is again largest for the Army, in which the coefficient suggests that hostile deployment reduces intention to reenlist by 10 percentage points. The evidence on

---

[1] The deployment variable comes from the administrative data, as do all of the deployment variables in our analysis (unless otherwise stated). Although the survey asks whether the respondent was "away" in the previous 12 months, we use the administrative data to indicate deployment because it differentiates between nonhostile and hostile deployment and may be more accurate. The deployment variables in the regressions are "nonhostile deployment only" and "hostile deployment"; "no deployment" is the base group. "Hostile deployment" signifies that the individual's deployment involved hostile duty, and it might have also included nonhostile deployment.

[2] These models also include controls for having only nonhostile deployment, spending more than one night away from home without being deployed, how prepared the respondent feels to carry out his or her job, AFQT category, location (rural or urban), education, race, marital status, whether the respondent is in a dual-service marriage, gender, survey wave indicator variables, one-digit MOS fixed effects, years of service, and pay grade. The regression results are reported in Appendix C.

Table 6.1
Estimated Hostile Deployment Effects on Work Stress, Personal Stress, Intention to Reenlist, and Reenlistment, First-Term Survey Respondents

Effect	Army	Navy	Marine Corps	Air Force
Higher-than-usual work stress	0.115**	0.075**	0.044*	−0.018
	(0.017)	(0.016)	(0.019)	(0.018)
Higher-than-usual personal stress	0.073**	0.065**	0.026	0.022
	(0.017)	(0.017)	(0.020)	(0.018)
Intention to reenlist	−0.102**	−0.044**	−0.095**	−0.062**
	(0.015)	(0.016)	(0.018)	(0.018)
Actual reenlistment	−0.056**	−0.012	−0.031	0.025
	(0.020)	(0.020)	(0.019)	(0.022)

NOTE: Cell entries are estimated regression coefficients on an indicator for hostile deployment in the 12 months prior to the survey. Robust standard errors are in parentheses. ** = statistically significant at the 1-percent level. * = statistically significant at the 5-percent level.

Table 6.2
Estimated Hostile Deployment Effects on Work Stress, Personal Stress, Intention to Reenlist, and Reenlistment, Second-Term-Plus Survey Respondents

Effect	Army	Navy	Marine Corps	Air Force
Higher-than-usual work stress	0.098**	0.126**	0.054**	0.020
	(0.012)	(0.013)	(0.019)	(0.014)
Higher-than-usual personal stress	0.093**	0.038**	0.037*	0.029*
	(0.012)	(0.013)	(0.018)	(0.014)
Intention to reenlist	−0.051**	0.021†	−0.009	−0.036**
	(0.011)	(0.011)	(0.016)	(0.012)
Actual reenlistment	0.008	0.118**	0.086**	0.028†
	(0.017)	(0.017)	(0.023)	(0.017)

NOTE: Cell entries are estimated regression coefficients on an indicator for hostile deployment in the 12 months prior to the survey. Robust standard errors are in parentheses.
** = statistically significant at the 1-percent level. * = statistically significant at the 5-percent level. † = statistically significant at the 10-percent level.

actual reenlistment is less clear. Unlike intention to stay, which is reported at the time of the survey, reenlistment typically occurs later, at the end of the term of service. For the Army, hostile deployment reduces reenlistment by about 6 percentage points. For the other branches, however, the effects are small in magnitude and not statistically significant.

The results for second-term-plus respondents (Table 6.2) also indicate that higher-than-usual work stress and higher-than-usual personal stress increase with hostile deployment, though in the Air Force, the effect on work stress is small and not statistically significant. As they were for first-term respondents, these effects are largest in the Army. The effects on intention to reenlist, however, are not as strong as they are for first-term respondents. In the

Army and, to a lesser extent, in the Air Force, hostile deployment is associated with reduced intention to reenlist, but in the Marine Corps, this relationship is very small and not statistically significant. In the Navy, it is actually positive though small in magnitude and only marginally significant.

The results suggest that hostile deployment is associated with an increase in actual reenlistment for all branches. This effect is strongest in the Navy and Marine Corps, in which hostile deployment increases reenlistment by 12 and 9 percentage points, respectively. A smaller effect (3 percentage points) is seen in the Air Force, and this coefficient is only marginally significant. The effect for the Army is less than 1 percentage point and not statistically significant.

One possible reason that the results for intentions and actual reenlistment differ is that intentions are measured long before the service members must actually make the decision and, in the intervening period, conditions changed that made them change their minds. To examine this possibility, we estimated models limiting the sample to respondents who had less than 18 months to ETS at the time of the survey. The results are generally similar for first-term respondents (the negative effect for the Army is slightly smaller in magnitude) and quite similar for second-term-plus respondents (the positive effect for the Marine Corps is somewhat larger). We also investigated the degree to which intentions predict actual behavior. Among respondents stating that they were not likely to reenlist and for whom a reenlistment outcome is observed, only 28 percent reenlisted. For those who said that they were likely to reenlist, about two-thirds actually did.

Estimates Controlling for Overtime Work and Deviations from Expected Time Away from Home

It is interesting to consider the mechanism driving these effects. One valuable feature of the survey data is information on overtime work and whether the service member has spent more or less time away from home than expected. Examining how the hostile deployment coefficients change when controlling for these variables provides insight into whether the effect of hostile deployment is "mediated" through these factors.

Tables 6.3 and 6.4 show the estimated deployment effects after controlling for overtime work and whether the respondent spent more or less time away than expected. For both first- and second-term-plus respondents, the addition of these covariates eliminates the positive association of hostile deployment on work and personal stress. In fact, for the Marine Corps and Air Force, conditional on the new covariates, hostile deployment actually has a *negative* association with work stress (though not with personal stress). Thus, it appears that hostile deployment does not contribute to increased levels of work and personal stress after accounting for the greater overtime work and unexpectedly long periods away from home brought about by deployments.

Interestingly, for first-term respondents, hostile deployment is still associated with lowered intention to reenlist in the Army and Marine Corps (and, to a lesser extent, in the Air Force) after controlling for these factors. For these respondents, there is something about hostile deployment that makes them less inclined to remain in the military above and beyond having to work harder and spending more time away from home. This is not the case for

Table 6.3
Estimated Hostile Deployment Effects, Controlling for Overtime Work and Time Away from Home, First-Term Survey Respondents

Effect	Army	Navy	Marine Corps	Air Force
Higher-than-usual work stress	0.010	0.010	−0.042*	−0.063**
	(0.018)	(0.017)	(0.020)	(0.019)
Higher-than-usual personal stress	0.017	0.024	−0.008	0.000
	(0.019)	(0.017)	(0.021)	(0.019)
Intention to reenlist	−0.051**	−0.000	−0.067**	−0.034[†]
	(0.017)	(0.017)	(0.019)	(0.019)
Actual reenlistment	−0.011	−0.002	−0.021	0.031
	(0.022)	(0.021)	(0.021)	(0.023)

NOTE: Cell entries are estimated regression coefficients on an indicator for hostile deployment in the 12 months prior to the survey. Regression models include controls for number of days working overtime and whether the respondent spent more or less time away from home than expected. Robust standard errors are in parentheses. ** = statistically significant at the 1-percent level. * = statistically significant at the 5-percent level. [†] = statistically significant at the 10-percent level.

Table 6.4
Estimated Hostile Deployment Effects, Controlling for Overtime Work and Time Away from Home, Second-Term-Plus Survey Respondents

Effect	Army	Navy	Marine Corps	Air Force
Higher-than-usual work stress	−0.025*	0.012	−0.072**	−0.070**
	(0.012)	(0.013)	(0.020)	(0.015)
Higher-than-usual personal stress	0.007	−0.014	−0.025	−0.017
	(0.013)	(0.014)	(0.020)	(0.015)
Intention to reenlist	−0.007	0.053**	0.026	0.003
	(0.012)	(0.012)	(0.017)	(0.013)
Actual reenlistment	0.014	0.100**	0.076**	0.027
	(0.019)	(0.018)	(0.024)	(0.018)

NOTE: Cell entries are estimated regression coefficients on an indicator for hostile deployment in the 12 months prior to the survey. Regression models include controls for number of days working overtime and whether the respondent spent more or less time away from home than expected. Robust standard errors are in parentheses. ** = statistically significant at the 1-percent level. * = statistically significant at the 5-percent level.

second-term-plus service members. For this group, the negative first-term associations observed for the Army, Marine Corps, and Air Force disappear when adjusting for the additional covariates. Further, the estimated effects on actual reenlistment for both first- and second-term-plus respondents generally do not change very much when adding the controls for overtime and time away from home. One exception is that the negative first-term effect for the Army (Table 6.1) essentially becomes zero. This suggests that the negative effect of hostile deployment on reenlistment may be driven by the work burden of deployment and spending more time away than was expected.

Estimates Using Similar Specification to That Used in the Administrative Data Analysis

The final set of analyses that we conducted with the survey data was to use a set of controls similar to that used in the administrative data analysis. Specifically, this specification excludes the controls for being away but not on deployment, job preparedness, location, marital status, and dual-service marriage. We also include three-digit DoD occupation codes and controls for the SRB multiplier in an occupation.[3]

Table 6.5 shows the results for first-term respondents. The effect of hostile deployment on work and personal stress generally follows the same pattern seen in the "baseline" estimates (Table 6.1), but the effects are smaller (negative and larger in magnitude for the Air Force). Similarly, the effects on intention to reenlist are also smaller, and the coefficient for the Navy is not statistically significant. The results for actual reenlistment are quite similar to the baseline estimates.

As was the case for first-term respondents, the estimated effects on work and personal stress for second-term-plus respondents are similar to the baseline estimates but smaller (the effect on personal stress for the Marine Corps is not statistically significant), as shown in Table 6.6. This is also the case for intention to reenlist. For reenlistment, the positive effects are slightly smaller and the modest effect for the Air Force is not significantly different from zero.

Table 6.5
Estimated Hostile Deployment Effects, Similar Specification as for Administrative Data, First-Term Survey Respondents

Effect	Army	Navy	Marine Corps	Air Force
Higher-than-usual work stress	0.072**	0.045**	0.034[†]	−0.036*
	(0.014)	(0.016)	(0.018)	(0.018)
Higher-than-usual personal stress	0.056**	0.054**	0.021	0.003
	(0.015)	(0.017)	(0.018)	(0.017)
Intention to reenlist	−0.067**	−0.009	−0.059**	−0.042*
	(0.013)	(0.016)	(0.016)	(0.017)
Actual reenlistment	−0.039*	0.009	−0.021	0.037[†]
	(0.017)	(0.020)	(0.017)	(0.021)

NOTE: Cell entries are estimated regression coefficients on an indicator for hostile deployment in the 12 months prior to the survey. Robust standard errors are in parentheses.
** = statistically significant at the 1-percent level. * = statistically significant at the 5-percent level. [†] = statistically significant at the 10-percent level.

---

[3] This is not identical to the specification used in the administrative data analysis. This specification includes pay grade controls, while the administrative data analysis controls for an indicator of promotion speed. We do not control for promotion speed because not all survey respondents had been in the service long enough to get to the important promotion points (i.e., to E-4 or E-5).

Table 6.6
Estimated Hostile Deployment Effects, Similar Specification as for Administrative Data, Second-Term-Plus Survey Respondents

Effect	Army	Navy	Marine Corps	Air Force
Higher-than-usual work stress	0.072**	0.105**	0.031[†]	0.003
	(0.010)	(0.013)	(0.017)	(0.014)
Higher-than-usual personal stress	0.077**	0.028*	0.022	0.011
	(0.010)	(0.013)	(0.017)	(0.013)
Intention to reenlist	−0.023*	0.029**	0.014	−0.021[†]
	(0.010)	(0.011)	(0.015)	(0.012)
Actual reenlistment	0.012	0.101**	0.086**	0.025
	(0.016)	(0.016)	(0.021)	(0.017)

NOTE: Cell entries are estimated regression coefficients on an indicator for hostile deployment in the 12 months prior to the survey. Robust standard errors are in parentheses.
** = statistically significant at the 1-percent level. * = statistically significant at the 5-percent level. [†] = statistically significant at the 10-percent level.

Conclusion

Our estimates using the survey data show that hostile deployment tends to increase work and personal stress and lower the intention to reenlist. Using the linked survey-administrative dataset, we find evidence of a negative deployment effect on reenlistment only for first-term Army respondents. All of these relationships are substantially weaker when we control for potential pathways through which hostile deployment might have an effect, such as days of overtime and deviations from expected time away from home. In contrast, the estimates change only moderately when using a specification similar to that used in the administrative data analysis described in the next chapter.

CHAPTER SEVEN

Empirical Results Using Administrative Data

This chapter presents our estimates of reenlistment models using administrative data from personnel and pay files. We first present overall results for the period from 2002 to 2007, broken out by term and service. Since the nature and intensity of hostile deployments change over this period, we then examine how the hostile deployment effects vary over time. The findings indicate that, in the Army, there was a sharp fall in the effect of hostile deployment on first- and second-term reenlistment, with the effect becoming negative in 2006. In the other services there was little deployment effect on first-term reenlistment; the effect on second-term reenlistment decreased until 2003 and then increased, remaining positive in all years. The chapter then looks more deeply at the Army and tries to uncover what lies behind the drop in the estimated deployment effects. In particular, we examine how the hostile deployment effects vary by deployment intensity as well as how they vary across subgroups.

Estimates of Deployment and Bonus Effects, 2002–2007

Deployment Effects by Service and First or Second Term

We begin with two tables showing the results based on decisions made after 2001, pooling across years and service-member subgroups. Table 7.1 shows the results from the first-term reenlistment model. The deployment variables in Table 7.1 refer to the 12-month window prior to the decision.[1] Each table has three panels corresponding to different specifications for the MOS controls. The first makes no adjustment for occupation, the second includes MOS fixed effects, and the third includes MOS fixed effects interacted with the quarter in which the decision was made. There are no estimates of the bonus effect from the third specification because the bonus does not vary within an occupation at a given point in time.[2] To be concise, only the estimates of the coefficients on deployment and the bonus multiplier are reported; coefficients for selected covariates are in Appendix C.[3]

---

[1] Appendix C reports similar results using a 36-month window.

[2] Bonuses can vary within an occupation for members in different service zones. However, virtually all individuals are in zone A (up to six years of service) at the time of their first reenlistment, and most are in zone B (7–10 years of service) at the time of their second reenlistment. Moreover, variation in bonuses across zones is not useful for identifying bonus effects because the bonus effect would be confounded with differences in years of service, which may be an indicator of tastes for military service.

[3] The models also include controls for years of service at the time of the decision, education, gender, AFQT category, race, an indicator for being promoted more rapidly than is typical, and year-of-decision indicators.

Table 7.1
Estimated Deployment Effects on Reenlistment, First-Term Decisions, 2002–2007

Deployment Effect	Army	Navy	Marine Corps	Air Force
No MOS controls				
Nonhostile deployment only	0.100**	0.107**	0.079**	0.184**
	(0.006)	(0.004)	(0.005)	(0.007)
Hostile deployment	−0.041**	−0.021**	−0.019**	−0.022**
	(0.003)	(0.003)	(0.003)	(0.003)
SRB multiplier	−0.009**	0.052**	0.035**	−0.007**
	(0.003)	(0.001)	(0.001)	(0.001)
MOS fixed effects				
Nonhostile deployment only	0.093**	0.104**	0.086**	0.177**
	(0.006)	(0.004)	(0.005)	(0.007)
Hostile deployment	−0.009**	−0.009**	0.015**	−0.004
	(0.003)	(0.003)	(0.003)	(0.004)
SRB multiplier	0.013**	0.065**	0.078**	0.011**
	(0.003)	(0.002)	(0.001)	(0.002)
MOS × quarter fixed effects				
Nonhostile deployment only	0.082**	0.097**	0.065**	0.164**
	(0.006)	(0.004)	(0.004)	(0.007)
Hostile deployment	−0.005[†]	−0.004	0.017**	0.000
	(0.003)	(0.003)	(0.002)	(0.004)

NOTE: Cell entries are estimated regression coefficients on an indicator for being deployed in the 12 months prior to the reenlistment decision. Robust standard errors are in parentheses. ** = statistically significant at the 1-percent level. [†] = statistically significant at the 10-percent level.

The results for first-term reenlistment indicate that deployment to a nonhostile location has a positive effect on reenlistment. This result holds across all three of the regression specifications and for each branch of service. In contrast, hostile deployments do not appear to substantially increase reenlistment. The specification with no MOS controls actually indicates that hostile deployments are associated with lower reenlistment. However, after controlling for occupation, the estimates suggest that any negative effect is quite small. In fact, for the Marine Corps, hostile deployments increase reenlistment by about 1.5 percentage points.

The deployment effect estimates that control for MOS fixed effects are nearly the same as those that control for MOS-by-quarter fixed effects. This is an important result because it suggests that any bias "spilling over" from bias in the bonus-effect estimate is small. Moreover, the pattern of results indicates that deployments are more likely in occupations with lower reenlistment rates but that this selectivity does not vary within an occupation over time.

Relative to the survey results, the most pronounced change is that the negative hostile deployment effect for the Army found for survey respondents is much smaller in the administrative data. There are several possible explanations. One is that many of the survey respondents made their decision in the later years of the 2002–2007 period (since there is a gap between the survey and the decision), but, as we discuss next, the effects for the Army fell dramatically during that period. A second possibility is that the survey respondents are systematically different from those who did not complete the survey questionnaire (i.e., nonrespondents). While this is possible, it should be noted that the confidence interval for the first-term Army coefficients includes the point estimates from the administrative data.

Table 7.2 shows analogous results for second-term reenlistment. As with first-term reenlistment, having only a nonhostile deployment has a large, positive effect on reenlistment. An important difference, however, is that hostile deployments have a positive and statistically significant effect on reenlistment across all branches.

The magnitude of the hostile deployment estimates increases when controlling for occupation, which, again, shows that hostile deployments are more likely in occupations that generally have lower reenlistment. The strength of this association is strongest in the

Table 7.2
Estimated Deployment Effects on Reenlistment, Second-Term Decisions, 2002–2007

Deployment Effect	Army	Navy	Marine Corps	Air Force
No MOS controls				
Nonhostile deployment only	0.102**	0.171**	0.115**	0.080**
	(0.005)	(0.005)	(0.008)	(0.007)
Hostile deployment	0.006*	0.039**	0.036**	0.018**
	(0.003)	(0.004)	(0.007)	(0.005)
SRB multiplier	0.007**	−0.016**	0.009*	−0.009**
	(0.003)	(0.001)	(0.004)	(0.001)
MOS fixed effects				
Nonhostile deployment only	0.102**	0.174**	0.126**	0.083**
	(0.005)	(0.005)	(0.008)	(0.007)
Hostile deployment	0.016**	0.042**	0.066**	0.027**
	(0.003)	(0.004)	(0.007)	(0.005)
SRB multiplier	0.025**	0.016**	0.004	0.019**
	(0.003)	(0.003)	(0.005)	(0.003)
MOS × quarter fixed effects				
Nonhostile deployment only	0.096**	0.169**	0.124**	0.080**
	(0.005)	(0.005)	(0.008)	(0.007)
Hostile deployment	0.014**	0.038**	0.072**	0.028**
	(0.003)	(0.004)	(0.007)	(0.005)

NOTE: Cell entries are estimated regression coefficients on an indicator for being deployed in the 12 months prior to the reenlistment decision. Robust standard errors are in parentheses. ** = statistically significant at the 1-percent level. * = statistically significant at the 5-percent level.

Marine Corps, in which a hostile deployment increases reenlistment by about 7 percentage points, and weakest in the Army.

Estimates of the Effect of Selective Reenlistment Bonuses on Retention

Tables 7.1 and 7.2 also show the estimates of the coefficient on the bonus multiplier. The results indicate that the estimated bonus effects are sensitive to the choice of controls for occupation, as expected from the discussion of bonus-setting behavior in Chapter Five and Appendix A. In the specification with no MOS fixed effects, the estimates are negative for the Army and the Air Force. Taken literally, this would mean that reenlistment would *decrease* in these services if bonus generosity increased. However, such a negative association is more likely attributable to larger bonuses being given in occupations that have more difficulty maintaining required staffing levels. When controlling for MOS fixed effects, the estimated bonus effects are positive and statistically significant for all branches on the bonus multiplier for the period 2002–2007. (Our preferred bonus estimates, discussed below, cover the period 1996–2007.)[4]

The magnitudes of these effects are substantial for first-term reenlistment in the Navy and Marine Corps. The estimates indicate that a one-unit increase in the SRB multiplier increases Navy reenlistment by 6.5 percentage points and Marine Corps reenlistment by 7.8 percentage points. These results may be compared to those of Hansen and Wenger (2002, 2005) and Hattiangadi et al. (2004). Hattiangadi et al. estimated a model of Marine Corps reenlistment from FY 1985 to FY 2003 and found that a one-step increase in the zone A (first-term) reenlistment bonus increased reenlistment by 6.6 percentage points, which compares with our 7.8 percentage points. Hansen and Wenger report that, during the period from FY 1987 to FY 2003, a one-step increase in the bonus was associated with an increase in Navy first-term reenlistment of 4.4 percentage points, which is lower than our estimate of 6.5 percentage points for the period 1996–2007. Appendix D discusses Hansen and Wenger's estimate versus ours.

Another issue to bear in mind when evaluating the bonus effects for the Navy is that the Navy was downsizing during this period. To help reduce the size of the force, the Navy may have cut or eliminated bonuses in occupations in which it also reduced the number of available reenlistment slots. This would generate a mechanical positive relationship between changes in the bonus and changes in reenlistment rates that does not reflect the supply response to a change in the bonus. Therefore, these estimates may provide a misleading sense of how reenlistment would change in a different context if bonuses were changed.

The bonus-effect estimates are much smaller for the Army and the Air Force. Similarly, the estimates for second-term reenlistment are also modest. The largest effect for second-term reenlistment is found for the Army, in which a one-unit increase in the bonus multiplier is associated with an increase in the likelihood of reenlistment of 2.5 percentage points. However, as noted in Chapter Five and Appendix A, even controlling for MOS fixed effects may not eliminate all endogeneity bias from the estimated bonus effects if changes within an occupation in terms of the size of the bonus are related to changes in the conditions determining reenlistment in an occupation. In particular, these results either are consistent with a smaller effect of the

---

[4]　We also attempted to estimate bonus effects that varied over time and that varied with deployment intensity (e.g., the number of months spent on deployment). However, the patterns were not conclusive and were sensitive to specification. Moreover, estimating and interpreting changes in bonus effects over time in a fixed-effects framework is complicated by the fact that the reasons that bonuses change within an occupation may be different in different periods. As a consequence, changes in the estimated effects may be due to differences in the extent of endogeneity bias rather than a true change in the effect of the bonus.

bonuses on second-term reenlistment, or the estimates for second-term reenlistment are more severely biased by endogeneity.

Deployment Effects on Retention Over Time

The pooled estimates reported in Tables 7.1 and 7.2 are instructive, but they do not take into account the changing nature of conflicts. As shown in Chapter Two, the casualty and death rate varied over time, and the cumulative time spent deployed increased dramatically after 2004. To investigate whether there were corresponding changes in the relationship between deployment and reenlistment, we estimated models separately by the year in which the decision was made. We also extended the estimation period back to 1996. To guard against any biases due to differential deployments across occupations or from bias in the bonus effect, tabled results from this point forward use the specification that controls for occupation-by-quarter fixed effects. We also discuss a specification in which the deployment effect is allowed to vary by year and bonus is included along with the other controls; this specification uses year and MOS fixed effects rather than occupation-by-quarter fixed effects. In Chapter Eight, we use the bonus estimate from this specification to see how much bonuses aided in counteracting the decreasing effect of deployment on Army reenlistment.

The panels in Figure 7.1 show for each service the estimated effect on reenlistment of having any hostile deployment (relative to no deployment of any type) in the 12 months prior to the decision. With a one-year window, the estimates can be interpreted as the effect of being deployed in the year before the decision was made. To make the figures more readable, only point estimates are displayed. The standard errors are generally small (less than 0.01) and are reported in Appendix C.

The results for the Army show that hostile deployment had a positive effect on reenlistment in the period prior to the start of the GWOT (i.e., prior to 2002). For first-term members, hostile deployment was associated with an increase in the reenlistment rate of a little more than 5 percentage points, and for second-term members, it was about 10 percentage points. Through 2004, the deployment effects remained positive and fairly stable, if a little smaller than in the pre-GWOT period.

This pattern abruptly changed in 2005, when the hostile deployment effect for second-term soldiers fell from 0.08 to near zero. In 2006, the effect on both first- and second-term reenlistment fell sharply to about −0.08. For first-term reenlistment, the estimate remained negative and substantial in magnitude in 2007 at about −0.05. For second-term reenlistment, the coefficient recovered in 2007 to about zero, though this was well below its earlier level. This pattern is especially striking given that deployment-related pay increased substantially between the 1996–2000 and 2002–2007 periods (Chapter Eight). Thus, hostile deployment depressed reenlistment in 2006 and 2007 despite the higher deployment pay.

In the Navy, hostile deployment does not appear to have an important effect on reenlistment for first-term service members. With the exception of 2001, when the estimated coefficient was −0.05, the deployment effects are small in magnitude and are not consistently positive or negative, nor is there a clear upward or downward trend. For the second-term decision, the estimates are typically positive, but there is a downward trend starting in 1998, with a bottoming out in 2003. This trend reversed, and, by 2005, the estimated effects were about 5–7 percentage points.

Figure 7.1
Estimated Effect of Hostile Deployment on Reenlistment for Hostile Deployment in 12 Months Prior to Reenlistment Decision, by Year of Decision and Service

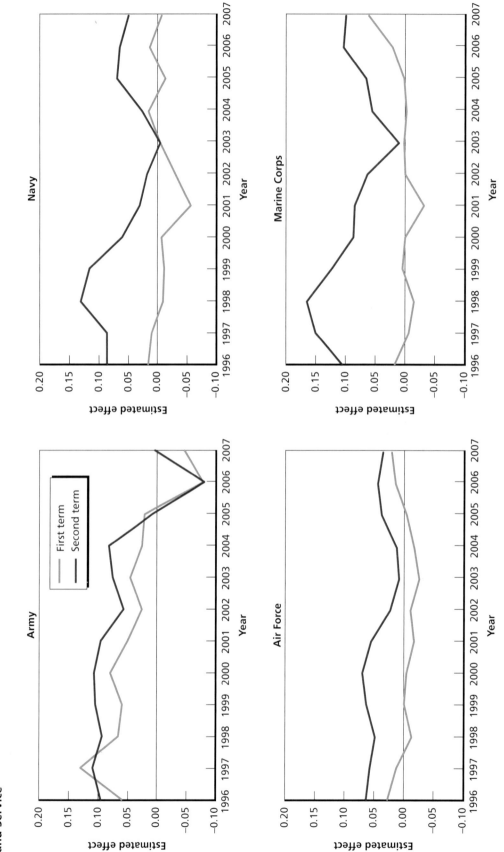

The Marine Corps results are similar to those for the Navy. Through 2005, the deployment effects for first-term marines are close to zero and exhibit no trend. In sharp contrast to the results for the Army, the deployment effect increases in 2006 and reaches 6 percentage points in 2007. The pattern for second-term reenlistment mirrors that of the Navy: a downward trend starting in 1998 that ends in 2003, followed by an upward trend.

The results for the Air Force show a stable pattern over time for the effect of deployment on first- and second-term reenlistment. Deployment has a larger (more positive) effect for second-term members, though the magnitude of the estimates is smaller than for the other services. As with the Navy and Marine Corps, the deployment effect for second-term decisions trends downward until 2003 and then upward, but the trends are less pronounced. For first-term reenlistment, the deployment effects are small in magnitude and fairly stable.

In summary, the effect of deployment on first-term reenlistment decreased in the Army and became negative, while the effect remained near zero in the other services. The effect of deployment on second-term Army reenlistment was much the same as for the first term. However, in the other services, the effect of deployment went down to 2003 and then increased, remaining positive in all years.

Did the Estimated Bonus Effect Also Change for the Army?

As noted earlier, the model used for Figure 7.1 that controls for occupation-by-quarter fixed effects cannot be used to estimate the effect of the bonus. Nonetheless, it is worthwhile to consider what the estimate of the bonus effect would be in the 1996–2007 period in a model that allows the deployment effect to vary by year. We ran such a specification (the results of which are presented in Table C.49 in Appendix C) that included the bonus multiplier and fixed effects by year and occupation (but not occupation by quarter). The deployment effects for the Army were close to those shown in Figure 7.1, but the bonus effect was 0.032 with a robust standard error of 0.002. This is larger than the bonus effect reported in Table 7.1, namely, 0.013, which is for a specification that constrains the deployment effect to be the same across years and is limited to the period 2002–2007. When the latter specification is estimated for 1996–2007, the bonus effect is 0.031 (0.002). The finding that the bonus effect over 1996–2007 is virtually the same whether or not the deployment effect is allowed to vary by year indicates that the correlation between deployment and bonus is quite low; this is borne out in auxiliary regressions (not reported). Further, the fact that the bonus effect is lower in 2002–2007 than for the entire period from 1996 to 2007 suggests that Army bonus setters managed bonuses more actively in 2002–2007, causing a greater downward bias. Although the estimate of 0.032 for the entire period contains this greater bias, it is probably a more accurate estimate of the bonus effect than the estimate of 0.013 for 2002–2007.

Another interesting question is whether the bonus effect changed in recent years. Starting in 2005, the Army began issuing bonuses in a lump-sum payment rather than half up front and the rest in equal annual installments. This may have changed the impact of a one-unit increase in the bonus multiplier. To investigate this possibility, we estimated the model but included an interaction between the bonus multiplier and an indicator for the 2005–2007 period. The estimated "main effect" for the bonus from this model is 0.040 (with a standard error equal to 0.002), and the interaction is –0.027 (with a standard error of 0.004), which suggests that the bonus effect in recent years is actually *smaller* (about 0.013) in the 2005–2007 period.[5] This is contrary to the prediction from the theory that a lump-sum bonus is more valuable than a bonus of equal nominal value paid over time.

Although we cannot determine why the bonus effects might differ over time, there are several possibilities. One is that the effect of the bonus truly is different in different periods, with a smaller effect in recent years. A second explanation is that the extent of the bias changed. If the Army began targeting bonuses more aggressively in the occupations that had the most pronounced reenlistment pressure in 2005–2007, then the downward bias would likely be worse.

Is the Army's Negative Deployment Effect in 2006–2007 the Result of Stop-Loss?

One possibility is that the negative effect of deployment on Army reenlistment in 2006 and 2007 comes from stop-loss. In general, stop-loss occurs when an individual is required to stay in the military beyond the ETS date. In the Army's application of stop-loss, a soldier whose term expiration date falls within 90 days of the unit deployment date can be stop-lossed. The stop-loss policy affects all soldiers who meet this criterion, but many of these soldiers are likely to reenlist. Stop-loss imposes a binding constraint only on those who want to exit. Stop-lossed individuals are allowed to exit only after the stop-loss directive is lifted. Thus, for these individuals stop-loss changes the timing of exit. It is also possible that the imposition of a stop-loss policy reduced the willingness to reenlist for individuals who were not stop-lossed simply because they did not like the idea that their choice to stay or leave might be constrained in the future if they were stop-lossed then.

In the context of our analysis, stop-loss pushes back the observed reenlistment decision date for individuals for whom stop-loss is binding. This could alter the composition of members making the decision in a particular year. For instance, if stop-loss forced individuals who would have exited in 2005 to delay until 2006, then our sample in 2006 would have a disproportionate number of exits.

We do not believe that stop-loss accounts for the drop in deployment effects in 2006 and 2007, for several reasons. First, stop-loss increased in the Army from November 2004 to March 2005, then decreased much of the way back to the 2004 level by January 2006 (Vanden Brook, 2008). Most of the soldiers stop-lossed in the early 2005 increase would have exited before 2006.[6] If so, a drop in deployment effects should have occurred in 2005, but the results are mixed. The effect on second-term reenlistment did decrease in 2005, but the effect on first-term reenlistment did not. By the same token, the compositional change associated with stop-loss should have led to an increase in the deployment effect in 2004. Our results show an increase in the effect for second- but not first-term reenlistment. Second, the stop-loss story would also imply that the counts of individuals making their reenlistment decision would change, with an increase in 2006 and a decrease in 2005. In contrast, we find that the number of first-term reenlistment decisions in our sample declined from 30,000 in 2005 to 27,000 in 2006.

---

[5]   The estimated deployment effect is 0.027 (with a standard error of 0.002). We also estimated a model that allowed the effect of hostile deployment to vary in each year and obtained similar bonus-effect estimates. The estimated deployment effects are similar to those reported in Figure 7.1.

[6]   In 2008, stop-lossed soldiers served an extra 6.6 months, on average (Vanden Brook, 2008). Applying a figure of six months' extra service to 2005, it is clear that most of the soldiers caught in the stop-loss increase in early 2005, and who wanted to separate, would have separated by the end of 2005.

Finally, and most importantly, we do not find any direct evidence that there was a higher fraction of constrained stop-lossed soldiers making decisions in 2006. Constrained soldiers are those who intend to exit but cannot because of stop-loss. Our data do not identify service members who were stop-lossed, but we were able to create an indicator of being constrained by stop-loss, which we defined as having the same value for months to ETS in consecutive months (i.e., the months-to-ETS countdown stops) in the 12-month period prior to reenlistment. If anything, stop-loss appears to be higher among individuals making a decision in 2004 or 2005 than in 2006, which is consistent with the information provided by Army officials. Table 7.3 shows our tabulations of the percentage of members stop-lossed, by service. The table includes percentages for 1996–2007. Our method is imperfect, and the earlier years may indicate possible false positives; we do not know the extent to which stop-loss was used before 2001. The percentages in the table are for the number of individuals at the reenlistment point who, by our measure, experienced a constraining stop-loss in the previous 12 months. The table shows an increase for the Army starting in 2003, and the percentages in 2006 and 2007 are lower than in 2004 and 2005.

In work related to ours, Simon and Warner (2009) estimate that soldiers covered under stop-loss were 24 percent less likely to reenlist in zone A (3–6 years of service) and 20 percent less likely to reenlist in zone B (7–10 years of service). By comparison, if we use our estimates of the percentage of soldiers constrained by stop-loss and assume that all of them went on to separate, we see a roughly 28-percent lower reenlistment rate at first term and 19-percent lower reenlistment at second term. These estimates are reasonably close to those of Simon and Warner. (Results available on request.)

On net, it seems unlikely that stop-loss is responsible for the negative effect of deployment on Army reenlistment in 2006 and 2007. This is not to overlook the possibility that the stop-

Table 7.3
Rough Estimate of Percentage Constrained by Stop-Loss, by Service, 1996–2007

Year	First Term				Second Term			
	Army	Navy	AF	MC	Army	Navy	AF	MC
1996	0.7	1.0	1.2	0.9	1.4	0.0	0.0	0.0
1997	0.7	1.5	2.2	0.7	0.5	1.3	13.3	1.1
1998	0.6	1.6	2.9	0.8	0.5	1.1	2.9	1.5
1999	0.5	1.4	2.6	1.2	0.5	1.5	2.2	1.2
2000	0.5	1.1	2.6	1.0	0.5	1.6	2.1	1.0
2001	0.7	0.6	2.8	1.0	0.7	1.0	5.3	1.3
2002	0.8	1.3	3.0	4.2	0.7	1.7	6.9	2.7
2003	2.8	1.3	4.2	2.1	1.9	1.9	4.9	1.7
2004	6.7	1.6	2.5	1.2	4.6	1.6	4.6	1.4
2005	10.0	1.5	2.8	1.2	6.1	1.7	4.0	1.3
2006	4.4	1.3	2.5	1.5	2.9	1.6	4.5	1.4
2007	3.6	1.3	2.9	1.5	2.8	1.9	3.5	1.2

loss policy, rather than actually being stop-lossed or constrained by stop-loss, accentuated the downward trend in the effect of deployment on Army reenlistment. That might have occurred, though we cannot identify it with our data; if an effect is present, it is embedded in the deployment effect for the stop-loss years.

Why Did the Deployment Effects Become Negative Only for the Army?

The divergence between the results for the Army and those for the other services seen in Figure 7.1 is clear. One potential source for this divergence is that there are differences in the extent of the deployment experience. Evidence in favor of this view can be seen in Figure 2.9 in Chapter Two, which shows that there was a striking increase in the fraction of members coming up for reenlistment who had 12 or more months of hostile deployment in the three years prior to the decision, coinciding with the fall in the estimated deployment effects. In contrast, there is no evidence of such movement for the Navy and Air Force. The Marine Corps represents a more interesting comparison, however, because, as in the Army, the fraction of service members with 12 or more months deployed increased substantially. By 2006, the Army had 52 percent with 12–17 months of hostile deployment in the three years preceding first-term reenlistment and 16 percent with 18 months or more. The corresponding figures for the Marine Corps are 30 percent and 5 percent, respectively. Whether these patterns can explain the divergence in the deployment effects depends on whether more extensive deployment has a larger negative effect on reenlistment than does less extensive deployment. The next section examines this proposition by comparing deployment effects for the Army and Marine Corps by the number of months deployed. Following those comparisons, we present results for service-member subgroups in the Army because that is where we find the most interesting changes in the aggregate estimates. Results for all branches are presented in Appendix C.

Hostile Deployment Effects by Extent of Deployment

To examine how the effects vary by the extent of deployment, we use three different measures of time spent deployed. All of these measures are based on a 36-month window prior to the decision because the shorter 12-month window would not be as useful for capturing multiple deployments or the total amount of time spent on deployment. The first approach uses variables based on the number of months on deployment: zero months, 1–6 months, 7–11 months, 12–17 months, and at least 18 months. The second approach breaks up the 36-month window into an "early" phase (the first 24 months) and a "late" phase (the last 12 months). Members are then classified into the following groups: those having any hostile deployment in both the prior 12 months and the first 24 months; those having hostile deployment in the prior 12 months only; those having hostile deployment in the first 24 months only; and not those deployed at all in the 36-month window.[7] Finally, variables are constructed on the basis of the number of unique deployment spells: none, exactly one, exactly two, and three or more.

Figure 7.2 shows the estimated effects by number of months deployed. The excluded category is no deployment of any kind in the 36-month window. The panels on the left show

---

[7] These groups refer to hostile deployments. The regression models include controls for being on a nonhostile deployment.

effects for the Army and the panels on the right show effects for the Marine Corps. For the Army, the effect of hostile deployment on reenlistment is higher for the low-months-deployed groups than for the high-months-deployed groups. This pattern becomes more pronounced over time, and by 2006 and 2007, having 1–6 months of hostile deployment is associated with a large increase in the reenlistment rate (15–25 percentage points), while having 18 or more months of deployment is associated with a large decrease in the reenlistment rate of more than 15 percentage points. The coefficients on all four months-deployed groupings decrease in 2006, but the sharp reduction in the overall deployment effect for the Army in 2006 (shown in Figure 7.1) appears to be driven by a decrease in the effect of being deployed 18 or more months or 7–11 months.

The results from the second-term reenlistment model (the lower panels in Figure 7.2) also show that the effect of deployment on Army reenlistment is lower for the high-months-deployed groups, particularly in the later years of the study period. Over time, the gap between the coefficients on the two lowest- and highest-month categories increases. By 2006, having 12 or more months of hostile deployment is seen to decrease reenlistment, while the effect for having fewer than 12 months increases it. The drop in the overall deployment effect in 2006 appears to be driven entirely by a steep fall in the coefficient on 18 or more months of hostile deployment, as the other coefficients do not change much between 2005 and 2006.

In contrast to those of the Army, the Marine Corps results are much noisier in the early years of the study period, when there are few marines with 12 or more months of deployment; this does not change until 2004. As a result, the deployment effects in the high-months-deployed groups are typically not statistically significant in the early years (standard errors are reported in Appendix C). In the low-months-deployed groups, we see that 1–6 months of deployment has no effect on first-term reenlistment except in the last few years, and 7–11 months of deployment has a negative effect in most years that averages about –0.05, i.e., first-term reenlistment is 5 percentage points lower among marines with 7–11 months of deployment versus marines with no deployment. This effect becomes positive in 2006 and 2007, however. In the high-months-deployed groups, the effect on first-term reenlistment is negative from 2004–2007 and often in the range of –0.05 to –0.10. The negative effect parallels the Army results but is more muted.

At second-term reenlistment, the effect of 1–6 months of deployment is positive throughout the study period and averages about 0.10. In fact, for most years, the effect of deployment on second-term reenlistment is positive for all months-deployed groups. The only notable negative effects are for the 18-or-more-months group, but these estimates are very noisy due to the small number of observations with 18 or more months deployed. Consequently, the effects for this group exhibit considerable year-to-year variation (climbing from –0.09 in 2004 to 0.09 in 2007).

Results based on our second approach are shown in Figure 7.3, which plots the estimated coefficients on the indicator variables for Army and Marine Corps deployment in the prior year only, in the first two years of the three-year window only, and in both the prior year and the first two years. The excluded category is no deployment of any kind in the three-year window, so the coefficients include the interpretation of the differential reenlistment rate associated with being in a given category relative to not having any hostile deployments. The results for first-term reenlistment are in the upper panels.

Figure 7.2
Hostile Deployment Effect by Months Deployed in 36 Months Preceding Reenlistment Decision, Army and Marine Corps

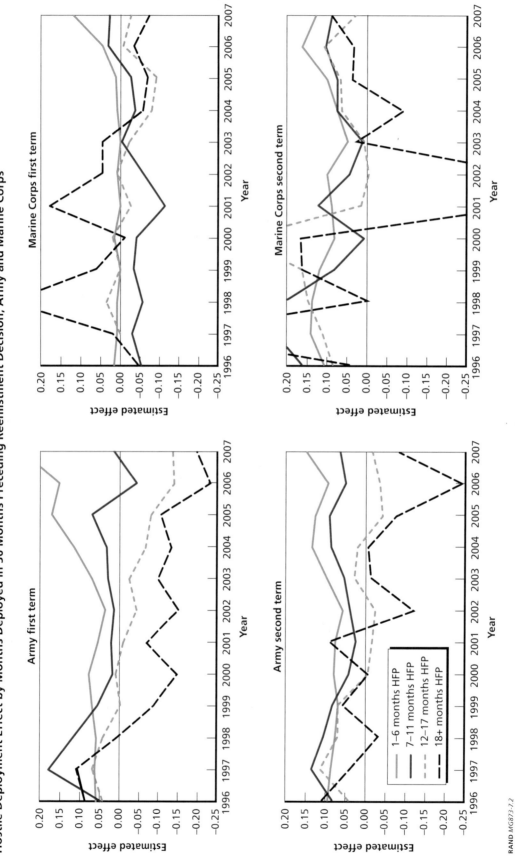

Figure 7.3
Estimated Effects of Hostile Deployment in Year Prior to Decision and in First Two Years of Window on Reenlistment, by Year of Decision, Army and Marine Corps

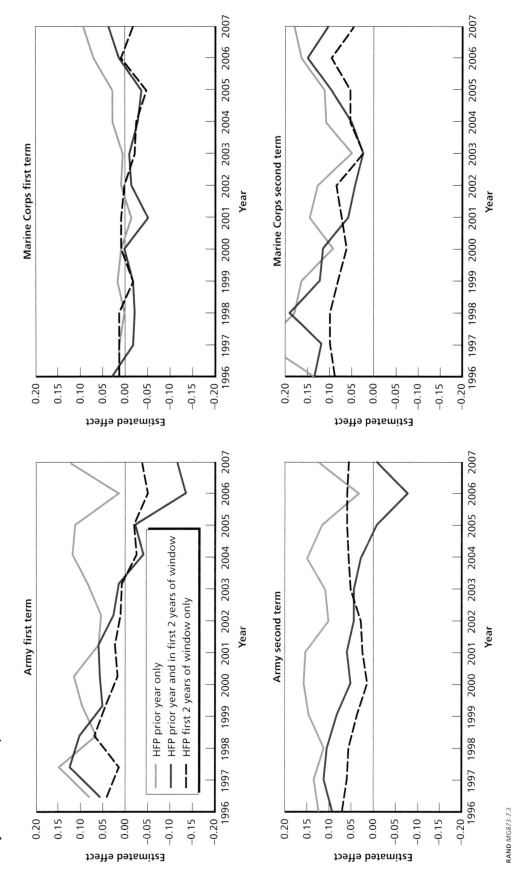

The Army coefficients for deployment in the prior year only do not appreciably trend upward or downward. The coefficient on a deployment only in the first two years of the window trends downward and becomes negative by 2004. The coefficient on deployment in both the prior year and in the first two years of the window also trends downward over time and, in 2006 and 2007, falls sharply and becomes negative and large in magnitude (less than –0.1). For second-term Army reenlistment (the lower left panel of Figure 7.3), the coefficients are typically larger (more positive) than they are for first-term reenlistment, with the only notable negative effect seen for deployment in both periods of the window in 2006.

Also noteworthy is that the rank ordering of the Army coefficients changed in recent years. Prior to 2004, having more deployment (as indicated by being deployed in both the prior year and the first two years) had a positive and larger coefficient than being deployed in the first two years only. In 2006 and 2007, however, being deployed in both the prior year and the first two years was associated with a very large reduction in the likelihood of reenlisting. These results suggest that having more extensive deployments in recent years is strongly associated with lower reenlistment, whereas this was not the case in earlier years.

The Marine Corps results (the right two panels) are easily summarized. The effect of deployment on first-term reenlistment is virtually zero from 1996 to 2002 for all three groups and then increases steadily for those with deployment in the prior year only. For the other two groups, both of which include deployment in the first two years of the three-year period before reenlistment, the effect remains close to zero through 2005 then turns up a bit in 2006 and 2007 (while remaining small in magnitude). At second-term reenlistment, the effect of deployment is positive and fairly similar for all three groups throughout the study period. In 2005–2007, the size of the effect is the same as in the first term: Deployment in the prior year only has the largest effect, followed by deployment in the first two years of the preceding three years, and then by deployment in both the prior year and the first two years.

Finally, we used the number of deployment episodes as another measure of the intensity of a service member's exposure to deployment. Unlike the findings for months deployed, the results for first- and second-term reenlistment for the Army and Marine Corps fail to show a clear relationship between the number of deployments and the likelihood of reenlisting (see Figure 7.4). Thus, to the extent to which greater deployment exposure has a "piling-on" effect that reduces reenlistment, it appears to be driven mainly through the number of months deployed rather than the number of separate deployment episodes. Because the number of months deployed is positively related to the number of deployments, it is difficult to disentangle the effects of total months deployed and number of deployments.[8] However, it may be that deployment months do a better job than do episodes at discriminating between deployments to Iraq and Afghanistan versus deployments to other areas deemed hostile but where deployments perhaps were shorter and less dangerous.

---

[8] To address this issue, we experimented with a specification that interacted the number of months deployed with the number of deployment episodes. The results suggested that having an additional deployment, conditional on months on deployment, is *positively* related to reenlistment. However, having an additional deployment conditional on months deployed is indicative of shorter deployment spells. Shorter deployments might be quite different from longer deployments in terms of the level of danger and the types of work that a service member is required to do. For example, Army deployments to Iraq have generally lasted 12 to 15 months, so a shorter deployment likely was to a different country.

Figure 7.4
Estimated Effects of Number of Hostile Deployments in Three Years Prior to Decision, Army and Marine Corps

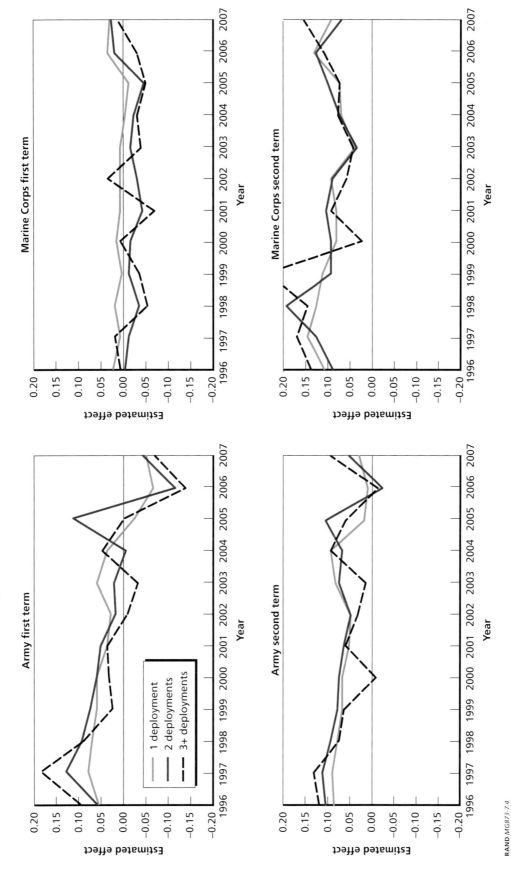

Effects by Service Member Subgroups

Examining results by subgroups is useful for two primary reasons. First, it offers additional insight into how effects vary by the nature of deployment. As noted earlier, our data provide little information on what happened during a deployment. We do, however, know that members serving in combat arms are at greater risk of injury or death than are members serving in different occupational specialties (this is also true for men compared to women). Examining results by these subgroups is therefore informative with regard to whether the change in the deployment effects seen in recent years for the Army is concentrated among members facing a disproportionately higher risk of physical harm. Second, it has been hypothesized that recent deployments have hit certain types of service members, such as those with families, especially hard. Comparing the deployment effects by marital status allows us to shed some light on this issue.

Figures 7.5 and 7.6 show the effects for men and women in the Army. In recent years, deployment appears to have had a larger downward effect (or a smaller positive effect) on reenlistment for men than for women, whereas this was not the case in earlier years. Figure 7.5 shows that the effects of deployment in the prior 12 months for first-term men and women were similar through 2005.[9] In 2006 and 2007, the coefficients for men are substantially lower than those for women. In 2007, the coefficient for women is essentially zero, while the coefficient for men remains negative and sizable. For second-term reenlistment, the coefficients for men and women are again similar through 2004, but then the deployment effect is much larger for women than for men.

Figures 7.7 and 7.8 show deployment effects by whether a member served in combat arms. For first-term reenlistment (Figure 7.7), the effect of deployment is larger among those serving in non–combat arms starting in 2000. This gap widens over time. By 2006, the effect of deployment is negative for both groups but is about 5 percentage points lower for members in combat arms. The patterns for second-term reenlistment (Figure 7.8) are similar, with the divergence between combat arms and non–combat arms starting two years later, in 2002.

Figures 7.9 and 7.10 show the deployment effects by marital status. Before 2001, the effect of hostile deployment was positive for both groups and larger for married service members. This difference was fairly stable through 2000. In 2001, the deployment effect among married service members began falling and eventually fell below that for singles. Although the size of the gap is fairly small (and switches sign in 2007), this is a notable change relative to the pattern seen before the start of the GWOT. In contrast, the effect of hostile deployment on second-term reenlistment (see Figure 7.10) is consistently larger for marrieds than for singles throughout the study period.

---

[9]   Similar results are obtained when examining deployment in a 36-month window.

**Figure 7.5
Estimated Effects of Hostile Deployment in Year Prior to Decision, by Gender, Army First Term**

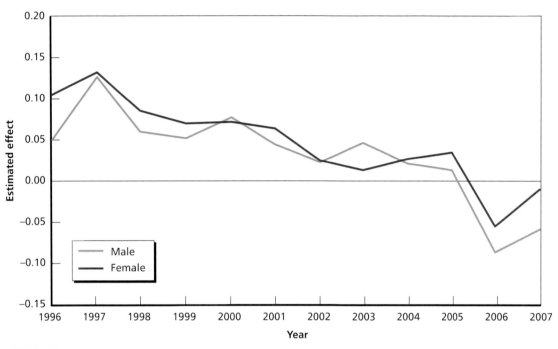

**Figure 7.6
Estimated Effects of Hostile Deployment in Year Prior to Decision, by Gender, Army Second Term**

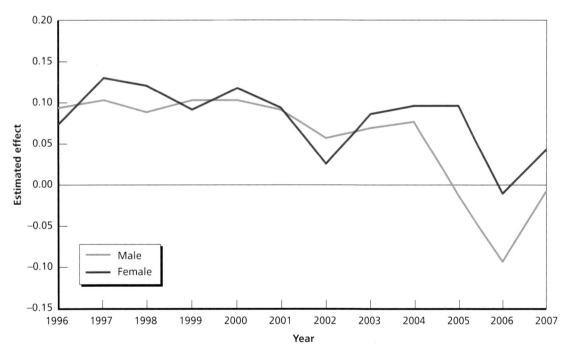

Figure 7.7
Estimated Effects of Hostile Deployment in Year Prior to Decision, by Whether Member Serves in Combat-Arms Occupation, Army First Term

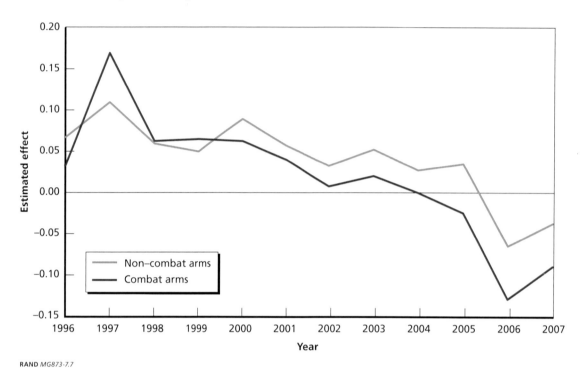

Figure 7.8
Estimated Effects of Hostile Deployment in Year Prior to Decision, by Whether Member Serves in Combat-Arms Occupation, Army Second Term

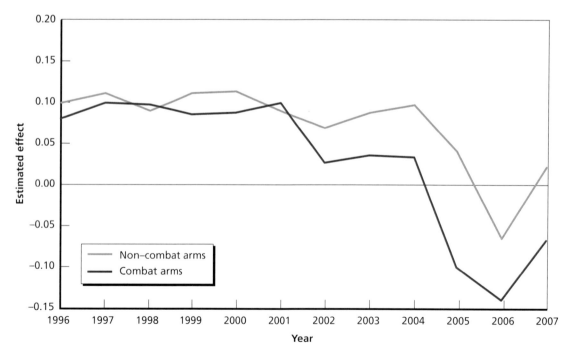

Figure 7.9
Estimated Effects of Hostile Deployment in Year Prior to Decision, by Marital Status, Army First Term

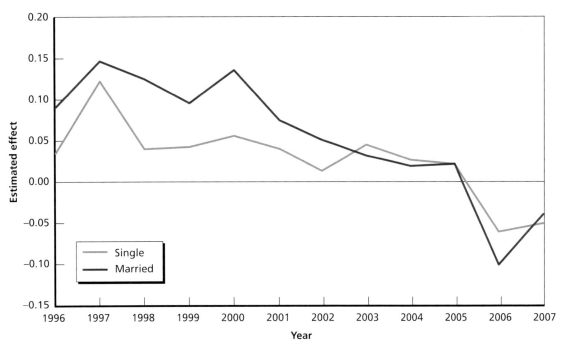

Figure 7.10
Estimated Effects of Hostile Deployment in Year Prior to Decision, by Marital Status, Army Second Term

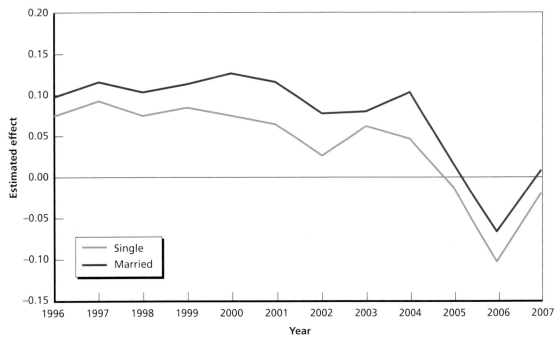

Conclusions

This chapter presents our estimates of the effect of deployment on reenlistment using data on the universe of reenlistment decisions made between 1996 and 2007 obtained from personnel records. It also presents estimates of the effect of bonuses on reenlistment, though these may be understated because of reverse causality generated by the fact that reenlistment bonuses are adjusted to offset unobserved (to the researcher) retention shocks or overstated if omitted variables on factors such as career-counselor effort or the availability of reenlistment slots are correlated with the bonus. The results indicate that hostile deployment generally had a positive effect on reenlistment over this period. However, in recent years, we find a striking divergence between the effects for the Army and for the other branches: The effect of hostile deployment for the Army became negative in 2006 and 2007. This effect was most pronounced among members serving in combat arms and felt most strongly among those individuals who had a relatively large number of months deployed. There is also evidence that a high number of months of deployment decreases the effect of hostile deployment on reenlistment in the Marine Corps; however, that effect does not turn negative. Our bonus estimates are positive for every service in models that control for occupation fixed effects, but we caution that the estimates might be biased.

CHAPTER EIGHT

The Role of Reenlistment Bonuses in Sustaining Retention

The purpose of this chapter is to discuss the role of bonuses in helping to sustain retention despite the downward effect of hostile deployment on Army first- and second-term reenlistment and on Marine Corps second-term reenlistment. This chapter builds on the model presented in Chapter Three. The model implies that, for any given level of deployment, it is possible to compute the bonus needed to restore the ex ante level of utility. For instance, suppose a soldier is thinking of reenlisting and initially expects for certain one hostile deployment of 12 months in a four-year term. Although the amount of time deployed is somewhat greater than the individual's preferred amount, which is 22 percent given the parameters of the model, assume that the individual is willing to reenlist. Given a change in expected deployment, we can calculate the bonus needed to restore utility to the initial value, which ensures that the individual will still reenlist. The change in expected deployment could be the result of new information about future deployment, e.g., how long the military operations will last and their size and pace.

Illustrative Examples

Table 8.1 shows the results of the bonus calculations for three different changes in expected deployment. As these cases show, both the expected amount of deployment and its uncertainty affect expected utility. First, suppose hostile deployment doubles but there is no uncertainty, i.e., suppose hostile deployment occurs half the time for certain rather than a fourth of the time for certain. A bonus of $0.11y$ is needed to restore expected utility. Second, suppose deployment becomes uncertain but its expected value remains the same; for instance, suppose the probability of zero deployment is one-fourth, the probability of quarter-time deployment is one-half, and the probability of half-time deployment is one-fourth. Then, a bonus of $0.19y$ restores utility. The third case combines greater deployment and uncertainty. It has a 50-percent chance of quarter-time deployment and a 50-percent chance of half-time deployment. Utility is restored by a bonus of $0.05y$.

The computations in Table 8.1 depend on the assumption about ex ante expected deployment. We chose a naïve assumption of quarter-time deployment with certainty, but better-informed expectations would allow for uncertainty, e.g., the chance of zero deployment, one deployment, or two or more deployments. First-term personnel might enter with naïve expectations and would form more sophisticated expectations at first-term reenlistment, having experienced deployment or heard about it from others. Second-term personnel would have better-informed expectations than would new enlistees and would continue to refine them.

Table 8.1
Deployment and Compensating Bonus, Given Initial Expectation of Hostile Deployment One-Fourth of the Time with Certainty

Case	Bonus to Restore Utility
Hostile deployment one-half of the time	0.11
0.25 chance of no deployment and 0.75 chance of deployment one-fourth of the time	0.16
0.50 chance of deployment one-half of the time and 0.50 chance of deployment one-fourth of the time	0.05
0.25 chance of no deployment, 0.50 chance of deployment one-fourth of the time, and 0.25 chance of deployment one-half of the time	0.19

NOTE: The bonus is expressed as a fraction of regular military compensation y. Ex ante utility assumes expected hostile deployment of 0.25 for certain.

Comparison to Deployment Pay and Potential Bonus Amounts

Although the bonus numbers in Table 8.1 are illustrative, it is useful to relate them to deployment pay and a typical service member's possible bonus at first-term reenlistment, say an E-4 in the fourth year of service. The two panels in Table 8.2 are used to derive deployment pay. The upper panel shows the components of monthly regular military compensation (RMC) for the period covered by our data, 1996–2008. The components are basic pay, basic allowance for subsistence, and basic allowance for housing, the amount of which depends on whether the service member has dependents. Basic pay is taxable, but the allowances are not taxable; thus, after-tax RMC is the sum of after-tax basic pay plus the two allowances. The table assumes a tax rate of 25 percent. The rightmost columns in the top panel express after-tax RMC in constant 2006 dollars. The lower panel shows the added compensation from serving on hostile deployment in a CZTE area, the components of which are the tax savings on basic pay plus HFP, hardship-duty pay, and, for members with dependents, family separation allowance. The rightmost columns show the added pay in 2006 dollars.

As seen in the upper panel, after-tax RMC in 2006 dollars increased from $1,793 in 1996 to $2,056 in 2008 for an E-4 without dependents and from $1,944 to $2,237 for an E-4 with dependents. These are increases of $263 and $293, respectively, and represent real pay raises of 15 percent. The lower panel shows that, over this period, monthly hostile deployment–related pay in 2006 dollars increased from $598 to $759 for an E-4 without dependents and from $695 to $993 for an E-4 with dependents. These are increases of $161 and $298, respectively, or 27 and 43 percent. The percentage increase in deployment pay is larger for the E-4 with dependents because of the large increase in the family separation allowance, from $75 in 1996 to $250 in 2003 and later.[1]

We now want to express hostile deployment–related pay in relative terms to associate it with the values used in the model. The second and third columns of Table 8.3 draw on the values in Table 8.2 to determine the ratio of hostile deployment–related pay to after-tax RMC. The next columns show the present value of a four-year reenlistment bonus with half paid up

---

[1] Table 8.2 does not include a bonus of up to $2,000 per year available to personnel who have completed a tour of duty and agree to extend the tour for at least one more year. The bonus was available for duty in Iraq and Afghanistan (Office of the Under Secretary of Defense [Comptroller], 2008).

Table 8.2
After-Tax Regular Military Compensation and Hostile Deployment–Related Pay for an E-4 in the Fourth Year of Service, Monthly, 1996–2008

Year	Basic Pay	Basic Allowance for Subsistence	Basic Allowance for Housing		After-Tax Pay		After-Tax Pay (2006 $)	
			w/o dep.	w dep.	w/o dep.	w dep.	w/o dep.	w dep.
1996	1,209	215	272	390	1,394	1,511	1,793	1,944
1997	1,246	221	285	408	1,440	1,563	1,810	1,965
1998	1,280	223	293	419	1,476	1,603	1,827	1,983
1999	1,327	225	304	434	1,524	1,654	1,845	2,003
2000	1,447	227	313	447	1,625	1,760	1,903	2,061
2001	1,501	230	323	463	1,679	1,818	1,912	2,071
2002	1,600	242	338	484	1,780	1,926	1,995	2,159
2003	1,665	243	353	505	1,844	1,996	2,021	2,188
2004	1,727	254	363	519	1,912	2,069	2,041	2,208
2005	1,787	267	379	542	1,986	2,150	2,050	2,218
2006	1,843	272	401	574	2,055	2,228	2,055	2,228
2007	1,883	280	415	594	2,107	2,287	2,050	2,224
2008	1,940	294	445	638	2,194	2,387	2,056	2,237

Year	Tax on Basic Pay Saved by CZTE	Hostile-Fire Pay	Hardship-Duty Pay	Family Sep. Allowance	Added Pay Related to Hostile Deployment		Added Pay Related to Hostile Deployment (2006 $)	
					w/o dep.	w dep.	w/o dep.	w dep.
1996	302	150	13	75	465	540	598	695
1997	311	150	13	75	474	549	596	691
1998	320	150	0	100	470	570	582	705
1999	332	150	0	100	482	582	583	704
2000	362	150	0	100	512	612	599	716
2001	375	150	100	100	625	725	712	826
2002	400	225	100	100	725	825	813	925
2003	416	225	100	250	741	991	812	1,086
2004	432	225	100	250	757	1,007	807	1,074
2005	447	225	100	250	772	1,022	796	1,054
2006	461	225	100	250	786	1,036	786	1,036
2007	471	225	100	250	796	1,046	774	1,017
2008	485	225	100	250	810	1,060	759	993

NOTE: 2006 dollar amounts are based on the consumer price index for urban consumers. Tax saved on basic pay is the tax saving on basic pay assuming that hostile deployment occurs in a CZTE area. Added pay related to hostile deployment without dependents is the sum of the tax excluded, HFP, and hardship-duty pay; with dependents, it is the same plus a family-separation allowance. These amounts are not taxed because they are received during deployment to a CZTE. The hardship-duty pay entries for 1996 and 1997 are based on foreign-duty pay, which was replaced by hardship-duty pay in 1998. HDP began in 2001 for Afghanistan and in 2003 for Iraq.

front and the remainder in annual installments discounted at 15 percent. The rightmost three columns show the present value of the bonus relative to four years of after-tax RMC. The deployment-related pay would have boosted the monthly after-tax RMC of an E-4 without dependents by a third in 1996 and by 37–41 percent since 2001. For an E-4 with dependents, the boost was 36 percent in 1996 and between 43 and 50 percent since 2001. (We use a value of 45 percent as a parameter in our computations with the utility model.) These are sizable increases in earnings over after-tax RMC. Given that our deployment effect estimates in the previous chapters included the effect of deployment pay, the fact that the hostile deployment effect for the Army became negative in 2005–2007 is all the more notable.

The middle columns of Table 8.3 show the present value of reenlistment bonuses at three different levels of generosity, step one, step two, and step three.[2] The bonuses are not taxed, which is consistent with assuming that reenlistment occurs in a CZTE area. If the bonuses

Table 8.3
Hostile-Deployment Pay and Selective Reenlistment Bonus Relative to After-Tax Regular Military Compensation for an E-4 in the Fourth Year of Service, Monthly, 1996–2008

Year	Hostile-Deployment Pay Relative to After-Tax Regular Military Compensation		Present Value of 4-Year Reenlistment Bonus			Bonus Relative to After-Tax Regular Military Compensation over 4 Years		
	w/o Dep.	w Dep.	Step 1	Step 2	Step 3	Step 1	Step 2	Step 3
1996	0.33	0.36	4,145	8,290	12,435	0.04	0.09	0.13
1997	0.33	0.35	4,269	8,539	12,808	0.05	0.09	0.14
1998	0.32	0.36	4,389	8,777	13,166	0.05	0.09	0.14
1999	0.32	0.35	4,547	9,094	13,641	0.05	0.09	0.14
2000	0.31	0.35	4,960	9,921	14,881	0.05	0.10	0.15
2001	0.37	0.40	5,143	10,287	15,430	0.05	0.10	0.16
2002	0.41	0.43	5,483	10,965	16,448	0.05	0.11	0.16
2003	0.40	0.50	5,708	11,416	17,123	0.05	0.11	0.16
2004	0.40	0.49	5,919	11,837	17,756	0.06	0.11	0.17
2005	0.39	0.48	6,125	12,251	18,376	0.06	0.12	0.17
2006	0.38	0.46	6,315	12,631	18,946	0.06	0.12	0.18
2007	0.38	0.46	6,454	12,909	19,363	0.06	0.12	0.18
2008	0.37	0.44	6,648	13,295	19,943	0.06	0.12	0.19

NOTE: Hostile-deployment pay relative to after-tax regular military compensation is based on the values in Table 8.2. The present value of the bonus assumes a four-year reenlistment; half of the bonus is paid immediately and the remainder in four installments, and the discount rate is assumed to be 0.15. Bonus relative to after-tax pay is the ratio of the present value of the bonus to 48 months of after-tax pay for an E-4 in the fourth year of service with dependents. The values are approximately 0.01 higher for personnel without dependents. An alternative approach of assuming basic pay based on future promotion and discounting future pay values produces similar results.

---

[2]   The reenlistment bonus amount equals monthly basic pay times years of reenlistment times bonus step. A step-three bonus is three times larger than a step-one bonus.

were taxed, the present values would be 25 percent lower at the assumed tax rate of 25 percent. The bonus tax saved from CZTE can be included in deployment-related compensation (though Table 8.2 does not do so). The rightmost columns show that adding or increasing a bonus by one, two, or three steps would increase four-year after-tax 2006 RMC by 6, 12, or 18 percent, respectively. Amounts for the years before 2006 are similar but slightly lower. These values correspond well with the bonus values in our example, shown in Table 8.1. That is, the bonuses were roughly the same as the bonus amounts computed in our example. This lends salience to the example and suggests that the bonuses that were actually paid may have been sufficient in many cases to restore a service member's initial level of utility and thereby induce the service member to reenlist despite expected deployment being much less than initially expected, much more than initially expected, or more uncertain than initially expected. Overall, the example illustrates the possibility that bonuses played an important role in sustaining reenlistment.

The next section provides information on bonus usage by service and reveals a large expansion of bonus allocation in the Army and Marine Corps during 2005–2007. Further, bonus generosity increased to levels consistent with those shown in our model and example.

Reenlistment Bonus Prevalence and Generosity

Figures 8.1–8.4 show bonus usage by service before and during the OEF/OIF years. Bonus usage is measured by the percentage of personnel reenlisting at the first term who received a bonus and the average bonus step. The Army and Marine Corps increased both the number and size of reenlistment bonuses in 2005–2007 compared with 2002–2004, and the increases were substantial. The percentage of soldiers reenlisting with a bonus decreased from 43–53 percent in 1999–2002 to 16 percent in 2003 and 2004, then increased to 71 percent in 2005 and 79 percent in 2006 and 2007 (see left panel of Figure 8.1). The large expansion in 2005 may have been driven by the Army's plans to increase personnel strength, enacted in 2004, rather than looming difficulties with reenlistment from deployment. The Army's average bonus step (for occupations that offered a bonus) stood at 1.6 in 2001 and 2002, fell to 1.3 in 2003 and 2004, and increased to 1.8 in 2005, 2.2 in 2006, and 2.1 in 2007 (see right panel in Figure 8.1). As a result of these changes, the percentage of reenlisting soldiers who received a bonus increased more than fourfold between 2003 and 2005–2007, and the average generosity of the bonus increased by more than 50 percent. The percentage of marines receiving a bonus increased from 43 percent in 2004 to 78 percent in 2007, and the average step climbed from 1.3 in 2002–2003 to 1.8 in 2004, 2.2–2.3 in 2005–2006, and 3.5 in 2007. In contrast, the percentage of sailors receiving a bonus held fairly steady at 70 to 80 percent, and the average step decreased from about 2.4 in 2003–2005 to 2.0 in 2006–2007. The Air Force decreased the percentage receiving a bonus from more than 80 percent in 2002–2003 to 14 percent in 2006–2007, and the average step increased from 3.2–3.4 in 2003–2004 to 3.7 in 2006–2007, a roughly 12-percent increase.

The evidence indicates that the branches that bore the brunt of the combat duties in OEF and OIF saw the largest increases in bonus usage and generosity. It is also true that the Army and Marine Corps embarked on efforts to grow in 2005. Interpreted in the context of our model, these bonuses compensated for the unexpected increase in deployment intensity experienced by service members in the Army and Marine Corps. Furthermore, the magnitude of the bonus increases is in line with what the model predicted would be needed to prevent a decline in reenlistment.

Figure 8.1
Reenlistment Bonus Prevalence and Average Step, Army First Term

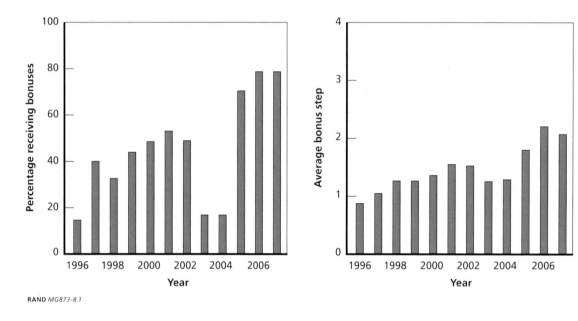

RAND *MG873-8.1*

Figure 8.2
Reenlistment Bonus Prevalence and Average Step, Marine Corps First Term

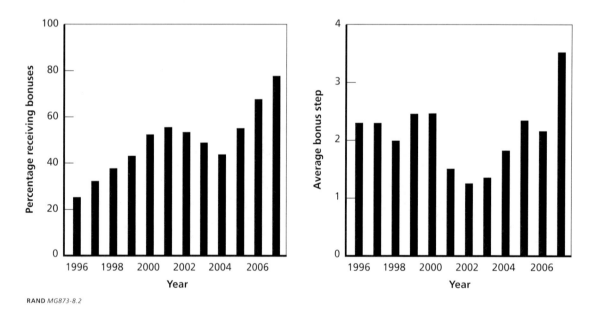

RAND *MG873-8.2*

Figure 8.3
Reenlistment Bonus Prevalence and Average Step, Navy First Term

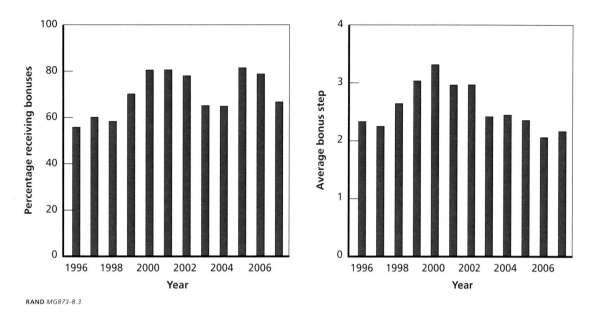

Figure 8.4
Reenlistment Bonus Prevalence and Average Step, Air Force First Term

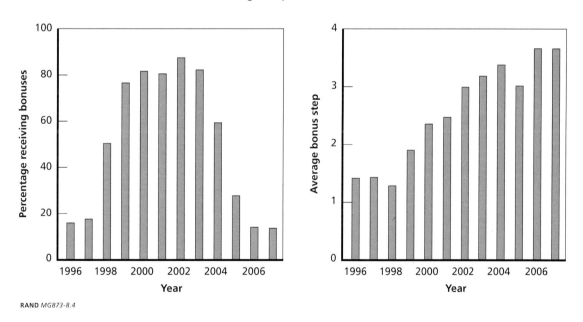

Accounting for the Impact of Bonuses on Army First-Term Reenlistment

The Army was hardest hit by the decreasing effect of deployment on reenlistment. The effect of deployment on Army first- and second-term reenlistment had become negative by 2006, and, as discussed, the Army greatly expanded its bonuses in 2005. Here, we demonstrate two accounting exercises to see whether the bonuses were sufficient to offset the downward trend in the effect of deployment, focusing on first-term reenlistment. The first exercise asks whether, in each year, the bonus was sufficient to offset the impact of deployment. This comparison is particularly useful for years in which the impact of deployment was negative—2006 and 2007. In other years, the impact of deployment was positive, and the impact of the bonus, also positive, adds further upward impetus to the reenlistment rate. The second exercise involves comparisons between selected years. This comparison reveals the change in reenlistment due to deployment, the change in reenlistment due to bonus, and whether the sum of these two changes was positive. Of particular interest are years (or a span of years) in which deployment reduced reenlistment, and the question is whether bonuses were sufficient to offset that reduction.

There are two basic parts to the first exercise: computing the impact of deployment and computing the impact of bonuses on the reenlistment rate. We use the regression specification with an indicator of hostile deployment in the previous 12 months. In our linear probability model, the impact of deployment equals the fraction of service members with deployment multiplied by the coefficient on deployment, and the impact of bonus equals the fraction receiving a bonus times the average bonus step times the coefficient on bonus. The sum of the deployment impact and bonus impact is the net effect on reenlistment. These quantities are given in Table 8.4.

The impact of deployment on reenlistment was positive and in the range of 0 to 3 percentage points from 1996 to 2005, but it fell to –5 percentage points in 2006 and to –3 percentage points in 2007. The impact of the bonus on reenlistment was positive and in the range of 0 to 2.7 percentage points from 1996 to 2004. With the sharp increase in bonuses in 2005, the bonus impact increased to 4.1 percentage points, and it increased further to 5.6 percentage

Table 8.4
Net Impact of Hostile Deployment and Bonus on Army First-Term Reenlistment, by Year

Year	Percent Deployed	Deployment Effect	Deployment Impact	Percent with Bonus	Average Bonus Step	Bonus Impact	Net Impact
1996	0.106	0.057	0.006	0.145	0.890	0.004	0.010
1997	0.168	0.117	0.020	0.399	1.049	0.013	0.033
1998	0.165	0.058	0.010	0.319	1.262	0.013	0.023
1999	0.158	0.058	0.009	0.435	1.260	0.018	0.027
2000	0.178	0.081	0.014	0.486	1.378	0.021	0.036
2001	0.145	0.068	0.010	0.530	1.572	0.027	0.036
2002	0.167	0.041	0.007	0.489	1.557	0.024	0.031
2003	0.347	0.021	0.007	0.163	1.257	0.007	0.014
2004	0.692	0.040	0.028	0.167	1.297	0.007	0.035
2005	0.611	0.021	0.013	0.708	1.832	0.041	0.054
2006	0.614	−0.083	−0.051	0.790	2.233	0.056	0.006
2007	0.620	−0.043	−0.027	0.787	2.098	0.053	0.026

NOTE: The bonus coefficient used in computing the bonus impact is 0.032. This is from a specification allowing the effect of deployment to vary by year but keeping the coefficients on other variables, including the bonus, the same in all years. Complete estimates are presented in Table C.49 in Appendix C.

points in 2006 and 5.3 percentage points in 2007. The net impact of deployment and bonuses was positive from 1996 to 2004 and in the range of 1 to 3.6 percentage points. This increased to a net effect of 5.4 percentage points in 2005 as the bonuses increased and the effect of deployment had not yet become negative; it then fell to 0.6 percentage points in 2006 and 2.6 percentage points in 2007. These results imply that the bonuses were sufficient to offset the negative effect of deployment.

For the second exercise, Table 8.5 shows the results of comparisons from one year to the next. Table 8.5 takes the deployment impact and the bonus impact from Table 8.4 and uses them to calculate the change in reenlistment owing to deployment and to bonuses. For instance, the impact of deployment on reenlistment was 0.010 in 2001 and 0.007 in 2002, resulting in a change in the impact of deployment of –0.003. Similarly, the change in the impact of the bonus was –0.002, and the sum of the changes was –0.005, or a 0.5-percentage-point decrease in reenlistment. The largest negative changes in the impact of deployment occurred in 2005 and 2006, with decreases in reenlistment of 1.5 and 6.4 percentage points, respectively. The expanded use of bonuses, which began in 2005, was sufficient to offset the impact of deployment in 2005 but not in 2006. Viewed from this perspective, bonuses helped to sustain reenlistment but were not enough to prevent a decrease. The fact that overall reenlistment did not decrease (see Figures 2.10, 2.11) must therefore be attributed to other positive changes in addition to the bonus. In our model, these factors are captured in the occupation-by-quarter fixed effects but are not identified. It is worth noting that the surge was not announced until January 2007, so it was not a factor in supporting reenlistment in 2006.

Table 8.5
Net Impact of Hostile Deployment and Bonus on Army First-Term Reenlistment, Between Years

| Year | Deployment Impact | Bonus Impact | Change in Reenlistment Due to | | Sum of Changes |
			Deployment	Bonus	
1996	0.006	0.004	—	—	—
1997	0.020	0.013	0.014	0.009	0.023
1998	0.010	0.013	–0.010	–0.001	–0.011
1999	0.009	0.018	0.000	0.005	0.004
2000	0.014	0.021	0.005	0.004	0.009
2001	0.010	0.027	–0.005	0.005	0.001
2002	0.007	0.024	–0.003	–0.002	–0.005
2003	0.007	0.007	0.000	–0.018	–0.017
2004	0.028	0.007	0.020	0.000	0.021
2005	0.013	0.042	–0.015	0.035	0.020
2006	–0.051	0.056	–0.064	0.015	–0.049
2007	–0.027	0.053	0.024	–0.004	0.021

NOTE: The bonus coefficient used in computing the bonus impact is 0.032. This is from a specification allowing the effect of deployment to vary by year but keeping the coefficients on other variables, including the bonus, the same in all years.

Conclusions

This chapter used the utility-maximizing model of deployment to show, under plausible parameters, that a compensating bonus can be paid to restore expected utility to its initial level. The bonus range suggested by the model corresponds to the actual range of bonuses paid by the Army and Marine Corps. The Army faced a decreasing effect of deployment on retention, and the bonuses helped to counteract this. In addition, both services had embarked on a program of growth in 2005, and the increased bonuses also helped to support growth by protecting reenlistment.

CHAPTER NINE

Conclusion

The burden placed on military personnel by OEF and OIF is undeniable. Before 9/11, the number of active-duty personnel receiving HFP was less than 50,000 per month. When U.S. forces fought Saddam's army in spring 2003, the number stood at 300,000 or nearly three in 10 of the 1.1 million-person active-duty force. In the years since then, 150,000 to 200,000 service members, mostly soldiers and marines, have received HFP each month. Since 2004, 80 percent of soldiers at reenlistment had been deployed at least once in the preceding three years, up from 30 percent in 1996–2002, and average months of hostile deployment had more than doubled.

The increase in deployments affected the services differently. In the Air Force, Navy, and Marine Corps, the effect of deployment on first-term reenlistment was at or near zero from 2002 to 2007. The Army's experience was quite different, however. The effect of deployment on Army reenlistment was positive in 1996–2001, decreased after 2002, and became negative in 2006. This was also true of Army second-term reenlistment. In the Navy and Marine Corps, the effect of deployment on second-term reenlistment decreased from a positive value in 1998 to zero in 2003, then rebounded. In the Air Force, the effect of deployment on second-term reenlistment remained positive throughout this period, declining slightly from 1998 to 2003, and then increasing. Although the Navy and Air Force were in the process of downsizing after 2003, this would not necessarily affect the estimate of the deployment effect. The Marine Corps had the same second-term pattern as the Navy, but the Marine Corps was in the process of growing.

We think that the expanded use of reenlistment bonuses was a major factor in why the increasingly downward pressure of deployment on Army reenlistment did not decrease the overall reenlistment rate. More than any other service, the Army increased the number of occupations eligible for a bonus and the amount of the bonus. The percentage of reenlisting soldiers who received a bonus increased from 15 percent in 2003–2004 to nearly 80 percent in 2005–2007, while the average value of bonuses increased by more than 50 percent. We do not have a definitive estimate of the bonus effect on reenlistment; estimation of the bonus effect is complicated by reverse causality, and our methods purge some but not all of the bias caused by this problem. However, our Army first-term bonus estimate was large enough in most years for the impact of bonuses to offset the downward pressure from deployment. Further, while our methods have controlled for some of the bias in the bonus-effect estimate, we think that a reenlistment bonus experiment (i.e., a randomized, controlled trial) will be necessary to obtain unbiased estimates of the bonus effect. Other factors in addition to bonuses may have been at work in sustaining reenlistment, such as improved communication links between deployed soldiers and their families back home, improved family support programs, and the Army's

Battlemind training program, which mentally prepares soldiers for what they will experience during deployment. Our data do not permit analysis of these factors.

The Army and Marine Corps experienced similar increases in deployments but managed their deployments differently. Marines typically had seven-month deployments to Iraq or Afghanistan, whereas soldiers had 12- to 15-month deployments. Our research does not address why the Army and Marine Corps chose different deployment lengths, and each service may have acted optimally given its roles and missions. What our findings do show is that the effect of deployment on Army reenlistment changed from positive to negative because more soldiers had high cumulative months of deployment and the estimated effect of deployment decreased from positive to negative for soldiers with many months of deployment. For marines, as mentioned, the effect of deployment on first-term reenlistment was zero, but *increased* and became positive after 2003, while the effect on second-term reenlistment declined, reached zero in 2003, and then rebounded. Note that these patterns mask heterogeneity because the the effect of having a high number of months deployed was negative for both the Army and Marine Corps. Nonetheless, *on average*, deployment did not negatively affect the reenlistment of marines because most marines who were deployed were not deployed for a large number of months.

The time-series of average months of deployment was similar for marines and soldiers at first-term reenlistment (see Figure 9.1) but was generally higher for soldiers, especially so in the early part of 2006, when the largest negative deployment effects for the Army are found. Moreover, the increase in the number of deployments was greater for the Marine Corps (see Figure 9.2). Given that Army deployments were 12–15 months and Marine Corps deployments were seven months, it may be worthwhile for the Army to consider shortening its deployments. In both the Army and Marine Corps, a large total number of months deployed reduced reenlistment in recent years, but this effect was larger for the Army. Further, an assessment of shorter versus longer deployments should consider mental health effects. Although overall reenlistment rates held steady and were supported by bonuses, it is possible that the prevalence of PTSD and major depressive disorder symptoms is higher among service members with longer deployments, holding total months of deployment constant. A higher prevalence of mental health symptoms, if present, would be a cost of longer deployment not reflected in reenlistment rates. Finally, perhaps the effect of deployment in the Army would be mitigated if the "dwell time" between deployments were lengthened, as opposed to deployments being shortened. These are questions for future study.

A novel feature of our study was the linking of survey data to administrative files on personnel and pay. The survey data consisted of 10 Status of Forces surveys of active-duty personnel between 2002 and 2005. The surveys contain information on intention to stay in the military as well as other variables not present in the administrative files. Results varied by service, but, broadly speaking, we found that the intention to stay was lower among those who had been deployed. Taken at face value, this is evidence that deployment would decrease retention. When we looked at actual reenlistment among the survey respondents, the effect of deployment was closer to zero, except for the Marine Corps. This suggests that intentions at the time of the survey had attenuated by the time of reenlistment. We did not detect that the attenuation was related to the number of months between the survey and the reenlistment decision. When we estimated models with the same explanatory variables but on the survey data versus the full administrative database, again with data pooled across years following the onset of the GWOT, we found little evidence that deployment reduced reenlistment. While

Figure 9.1
Average Months of Hostile Deployment Over 36 Months Preceding First-Term Reenlistment Decision, Army and Marine Corps

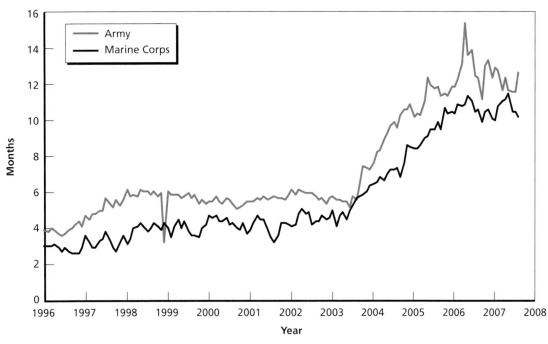

Figure 9.2
Average Number of Hostile Deployments Over 36 Months Preceding First-Term Reenlistment Decision, Army and Marine Corps

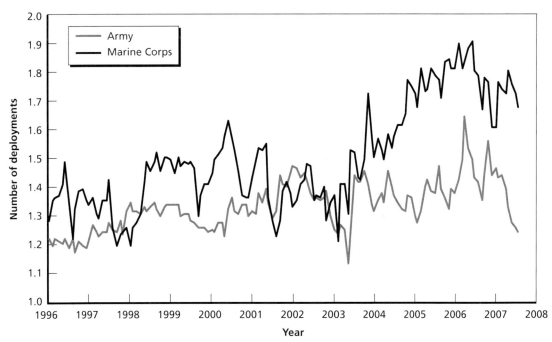

there was no systemic pattern across the services at first term between the deployment effect estimates for survey respondents and those for the rest of the population, at second term, the deployment effect estimates for all services were lower for survey respondents. This suggests that survey respondents at second term were biased toward a negative effect of deployment on reenlistment.

This and previous studies have found that, in most instances, deployment increased reenlistment. More precisely, having some deployment increased reenlistment, but extensive deployment can decrease reenlistment. In the Navy, Air Force, and Marine Corps, the effect of deployment was zero or positive in 2002–2007. In the Army, it changed from positive to negative, with the negative effects occurring among soldiers with a high number of cumulative months of deployment. The finding that some deployment increases reenlistment while too much can decrease it is consistent with the view that deployment is a direct source of utility, as modeled in Chapter Three. Deployment is not simply another form of "work" that takes time away from "leisure" and monotonically reduces utility, as in the traditional labor supply model. This pattern is also consistent with the fact that service members receive higher pay while deployed, and the higher pay helps to offset the negatives of hostile deployment.

A Model of Reenlistment Bonus Setting

This appendix presents a model of bonus-setting behavior that formalizes the discussion in Chapter Five that the endogeneity of reenlistment bonuses can bias bonus-effect estimates. The model assumes that the military sets bonuses with the aim of reaching a target reenlistment rate. We consider a constant target rate, an updated target rate, and a noisy target rate. Most elements of military compensation are taken as given by the services, at least in the short run, but bonuses are under the military's control. Legislation establishes parameters for when a bonus may be paid, the formula for the bonus amount, and the maximum payable, and the military bonus setter determines which specialties receive a bonus and its generosity. Because bonuses are set by the military, the bonus-setting process should be considered when determining how to specify or interpret a reenlistment model that includes a bonus as an explanatory variable.

Although our empirical analysis is based on micro data, it is useful to begin at the occupational level because this is where bonus setting takes place. The final section of this appendix presents a micro-level model of reenlistment.

Occupation-Level Model

We show that when autocorrelation is present, the bonus effect in a reenlistment equation may be biased downward. The bias is unambiguous when the reenlistment target is constant but can move in either direction when the target is updated. The analysis also demonstrates the importance of controlling for fixed effects by MOS. When fixed effects are not controlled, the bonus-effect estimate combines the true effect and fixed effect. We assume that autocorrelation is first-order; the current error depends on last period's error plus an uncorrelated random draw and does not depend on errors before the last period. Extending to higher-order autocorrelation would produce similar insights.

Constant Target
The structural reenlistment equation for an occupation is

$$r_{ct} = \beta_0 + \beta_1 B_{ct} + \mu_c + \varepsilon_{ct}.$$

The reenlistment rate here depends four factors: a constant, the bonus, a fixed effect, and an error term. Other variables, such as the unemployment rate, are assumed for simplicity to be in the constant.[1] The bonus effect on the reenlistment rate is assumed to be positive $\left(\beta_1 > 0\right)$ and the same for all occupations. The term μ_c is a fixed effect for the military occupation and represents such factors as work conditions, the value and transferability of training, exposure to risk, and civilian job opportunities that persistently differentiate this military specialty from others. The error is autoregressive:

$$\varepsilon_{ct} = \rho \varepsilon_{ct-1} + \upsilon_{ct}.$$

We assume that occupations follow the same autoregressive process (ρ is the same across occupations). The errors, ε_{ct}, are not correlated across occupations in the same period or between different periods; the same holds for the random draws, υ_{ct}, and they are not correlated between periods in a given occupation. The υ_{ct} have mean zero and variance σ^2. The reenlistment target, $\bar{r}_c$, is assumed to be constant.

The bonus setter observes the reenlistment rate and bonus in period $t-1$ and knows the parameters $\left(\beta_0, \beta_1, \mu_c, \rho\right)$. The bonus for period t is set so that the expected value of the reenlistment rate equals the target rate. The expected value of the reenlistment rate for some arbitrary bonus equals

$$Er_{ct} = \beta_0 + \beta_1 B_{ct} + \mu_c + \rho \varepsilon_{ct-1}.$$

The expected value incorporates the information from the error term in the previous period.[2] Using $Er_{ct} = \bar{r}_c$ and solving, the bonus is set at

$$B_{ct} = \frac{\left(\bar{r}_c - \left(\beta_0 + \mu_c + \rho \varepsilon_{ct-1}\right)\right)}{\beta_1}.$$

A higher target implies a higher bonus, whereas an attractive military occupation, $\left(\mu_c > 0\right)$, calls for a lower bonus. When the bonus is set by this method, it will be correlated with the error term. In particular, the covariance between the bonus and the error in period t is

$$\operatorname{cov}\left(\frac{\bar{r}_c - \left(\beta_0 + \mu_c + \rho \varepsilon_{ct-1}\right)}{\beta_1}, \varepsilon_{ct}\right) = -\frac{\rho^2}{1-\rho^2}\frac{\sigma^2}{\beta_1}.$$

The derivation of this result uses the recursive nature of the errors.[3] The negative covariance implies a downward bias in the estimate of the bonus effect from a regression of the retention rate on the bonus and occupation effects. The bias will be larger if the period-to-period error correlation is larger. If there is no autocorrelation, the bonus effect will not be biased.

---

[1]   Alternatively, time-varying factors that affect all occupations could be represented by a period-specific intercept.

[2]   If the bonus setter knows μ_c, then the random term from the prior period can be backed out once the reenlistment rate from that period is observed. With a long time series in a stable environment, it is reasonable to assume the bonus setter knows the occupation-specific fixed effect.

[3] $\operatorname{cov}\left(\rho \varepsilon_{ct-1}, \varepsilon_t\right) = \operatorname{cov}\left(\rho \varepsilon_{ct-1}, \rho \varepsilon_{ct-1} + \upsilon_{ct}\right) = \rho^2 \operatorname{cov}\left(\varepsilon_{ct-1}, \varepsilon_{ct-1}\right)$. Also,

$\varepsilon_{ct-1} = \rho \varepsilon_{ct-2} + \upsilon_{ct-1} = \rho \left(\rho \varepsilon_{ct-3} + \upsilon_{ct-2}\right) + \upsilon_{ct-1} = \rho^2 \left(\rho \varepsilon_{ct-4} + \upsilon_{ct-3}\right) + \rho \upsilon_{ct-2} + \upsilon_{ct-1} = \rho^{k-1} \varepsilon_{t-k} + \sum_{j=1}^{k-1} \rho^{j-1} \upsilon_{t-j}.$

Following this thought, we can see how the estimate of the bonus effect, β_1, would change if occupation fixed effects were omitted from the estimation equation. In that case, the error becomes $\mu_c + \varepsilon_{ct}$, and the covariance between the bonus and the random term is

$$\text{cov}\left(\frac{\bar{r}_c - \left(\beta_0 + \mu_c + \rho\varepsilon_{ct-1}\right)}{\beta_1}, \varepsilon_{ct} + \mu_c\right) = -\frac{1}{\beta_1}\left(\text{cov}\left(\rho\varepsilon_{ct-1}, \varepsilon_{ct}\right) + \text{var}\left(\mu_c\right)\right)$$

$$= -\frac{1}{\beta_1}\left(\frac{\rho^2}{1-\rho^2}\sigma^2 + \text{var}\left(\mu_c\right)\right).$$

Now, the covariance between the bonus and the random term is larger, resulting in a larger downward bias.

Suppose autocorrelation is absent. Given that the reenlistment equation includes an intercept and a fixed effect for occupation, variation in the bonus comes from variation in the target reenlistment rate. Thus, the bonus effect is identified in this non-autocorrelation model by exogenous variation in the reenlistment target across occupational specialties, given the assumption that the bonus effect is the same across occupations. As a related point, if the model does not include occupation fixed effects or imperfectly controls for occupational fixed effects by grouping occupations as some studies have done (see, e.g., the summary in Goldberg, 2001), exogenous variation in the bonus comes from $\bar{r}_c - \mu_c$, i.e., from both the target and the occupation fixed effect, and this difference can be positively or negatively related to the dependent variable r_{ct}, producing either a positive or negative "effect" of the bonus. Consider two extremes. If the retention target is the same across occupations $\left(\bar{r}_c = \bar{r}\right)$, then the bonus would be higher in less attractive occupations. In our model, the bonus would be set just high enough to offset the negative features of the occupation, and a higher bonus would not be associated with a higher reenlistment rate. In fact, the reenlistment rate would be the same in expectation across occupations, and the estimated bonus effect would be zero. If the bonus did not fully compensate for the occupation's unattractive features, a higher bonus would be associated with a lower reenlistment rate, i.e., the estimated bonus effect would be negative. At another extreme, suppose occupations are equally attractive (have the same fixed effect) but the target varies across occupations. Occupations with the highest target would have the highest bonus, and the estimated bonus effect would be positive. It would also be unbiased, assuming no autocorrelation.

Updated Target

The target is updated to increase if the reenlistment outcome in period t–1 was below the long-term target and to decrease if it was above the long-term target. For example, consider partial updating, i.e., a lagged adjustment model:

$$\bar{r}_{ct} = r_{ct-1} + \gamma\left(\bar{r}_c - r_{ct-1}\right) = \gamma\bar{r}_c + \left(1-\gamma\right)r_{ct-1} \qquad 0 \leq \gamma \leq 1.$$

Therefore, $\varepsilon_{ct-1} = \sum_{j=1}^{\infty}\rho^{j-1}v_{t-j}$. So, $\text{cov}\left(\varepsilon_{ct-1}, \varepsilon_{ct-1}\right) = \sigma^2 + \rho^2\sigma^2 + \rho^4\sigma^2 + \dots = \frac{1}{1-\rho^2}\sigma^2$.
$\lim_{k\to\infty}$

An unexpectedly low reenlistment rate in the previous period leads to a desired reenlistment rate in the current period that is below the long-run target. Such a model might reflect costly adjustments that make it inefficient to reach the long-term target immediately. This model might not be particularly relevant to the military, in which the cost of not meeting a reenlistment target may be extremely high, resulting in understaffed units and the misallocation of personnel, e.g., recently promoted sergeants with little leadership experience assigned to train privates. Using the previous approach, the bonus is set at

$$B_{ct} = \frac{\left(\bar{r}_{ct} - \left(\beta_0 + \mu_c + \rho\varepsilon_{ct-1}\right)\right)}{\beta_1}.$$

Substituting in for $\bar{r}_{ct}$:

$$B_{ct} = \frac{\left(\gamma\bar{r}_c + \left(1-\gamma\right)r_{ct-1} - \left(\beta_0 + \mu_c + \rho\varepsilon_{ct-1}\right)\right)}{\beta_1}$$

$$= \frac{\left(\gamma\bar{r}_c + \left(1-\gamma\right)\left(\beta_0 + \beta_1 B_{ct-1} + \mu_c + \varepsilon_{ct-1}\right) - \left(\beta_0 + \mu_c + \rho\varepsilon_{ct-1}\right)\right)}{\beta_1}$$

$$= \frac{\left(\gamma\left(\bar{r}_c - \beta_0 - \mu_c\right) + \left(1-\gamma\right)\beta_1 B_{ct-1} + \left(1-\gamma-\rho\right)\varepsilon_{ct-1}\right)}{\beta_1}.$$

Substituting in for B_{ct-1} and proceeding by recursion, the bonus in period t equals a constant term, $\left(\bar{r}_c - \beta_0 - \mu_c\right)/\beta_1$, plus terms reflecting the influence of past error terms:

$$B_{ct} = \frac{\left(\begin{array}{l}\gamma\left(\bar{r}_c - \beta_0 - \mu_c\right)\left(1 + \left(1-\gamma\right) + \ldots + \left(1-\gamma\right)^{j-1}\right) + \left(1-\gamma\right)^j \beta_1 B_{ct-j} \\ + \left(1-\gamma-\rho\right)\left(\varepsilon_{ct-1} + \left(1-\gamma\right)\varepsilon_{ct-2} + \ldots + \left(1-\gamma\right)^{j-1}\varepsilon_{ct-j}\right)\end{array}\right)}{\beta_1}$$

$$B_{ct}_{\lim j\to\infty} = \frac{\left(\left(\bar{r}_c - \beta_0 - \mu_c\right) + \left(1-\gamma-\rho\right)\left(\varepsilon_{ct-1} + \left(1-\gamma\right)\varepsilon_{ct-2} + \ldots + \left(1-\gamma\right)^{j-1}\varepsilon_{ct-j} + \ldots\right)\right)}{\beta_1}.$$

The constant term can be thought of as the long-term or steady-state bonus that would be expected to produce the target reenlistment rate. That is, the expected value of the reenlistment rate under the long-term bonus taken over all possible values of the error term is $\bar{r}_c$:

$$Er_{ct} = E\left(\beta_0 + \beta_1\left(\frac{\left(\bar{r}_c - \beta_0 - \mu_c\right)}{\beta_1}\right) + \mu_c + \varepsilon_{ct}\right) = \bar{r}_c.$$

Using the formula for B_{ct}, the covariance between the bonus and the error is as follows:[4]

$$\mathrm{cov}\left(\frac{\left(\left(\bar{r}_c - \beta_0 - \mu_c\right) + \left(1 - \gamma - \rho\right)\left(\varepsilon_{ct-1} + \left(1 - \gamma\right)\varepsilon_{ct-2} + \ldots + \left(1 - \gamma\right)^{j-1}\varepsilon_{ct-j} + \ldots\right)\right)}{\beta_1}, \varepsilon_{ct} \right)$$

$$= \frac{\left(1 - \gamma - \rho\right)}{\beta_1}\frac{\rho}{1 - \rho^2}\frac{\sigma^2}{1 - \left(1 - \gamma\right)\rho}.$$

The sign of the covariance depends on the sign of $\left(1 - \gamma\right) - \rho$, where $\left(1 - \gamma\right)$ is the weight on r_{ct-1} in setting the target for period t, and ρ is the period-to-period error correlation. If the weight placed on the reenlistment outcome in the previous period—$\left(1 - \gamma\right)$—is small and the error correlation is large and positive, then $\left(1 - \gamma\right) - \rho$ will be negative. As before, this will result in a downward bias on the bonus effect. But if the error correlation is a small positive number and high weight is placed on the reenlistment outcome in the previous period, the bias is upward. Also, if the correlation is negative, the bias is upward. In the case where $\left(1 - \gamma\right)$ equals ρ, there is no bias on the bonus effect.

On net, the lagged value of r_{ct-1} used in setting the target induces a positive covariance between the bonus and the error in period t, whereas the autocorrelated error induces a negative covariance. The intuition for the first statement is as follows. When retention is below the long-run target in the previous period, the bonus setter revises the desired rate to be below the long-term rate (perhaps because of adjustment costs). The bonus setter therefore sets a lower bonus than would otherwise be the case in the current period. If the negative shock that led to low reenlistment persists (as would be true with positive autocorrelation), then the target updating process would introduce a positive association between the bonus and the random term that offsets (and potentially overturns) the negative association between the bonus and the random term that, as we saw before, is a direct consequence of positive autocorrelation.[5]

---

[4]   The covariance expression expands to a weighted sum of the covariance between the error term in period t and the error term in each earlier period:

$$\frac{\left(1 - \gamma - \rho\right)}{\beta_1}\left(\mathrm{cov}\left(\varepsilon_{ct-1}, \varepsilon_{ct}\right) + \left(1 - \gamma\right)\mathrm{cov}\left(\varepsilon_{ct-2}, \varepsilon_{ct}\right) + \ldots + \left(1 - \gamma\right)^{j-1}\mathrm{cov}\left(\varepsilon_{ct-j}, \varepsilon_{ct}\right) + \ldots\right).$$

Taking each covariance term separately, we used the recursive approach described in footnote 3.

$$\mathrm{cov}\left(\varepsilon_{ct-1}, \varepsilon_{ct}\right) = \mathrm{cov}\left(\varepsilon_{ct-1}, \rho\varepsilon_{ct-1} + \upsilon_{ct}\right) = \rho\,\mathrm{cov}\left(\varepsilon_{ct-1}, \varepsilon_{ct-1}\right) = \frac{\sigma^2\rho}{\left(1 - \rho^2\right)}$$

$$\mathrm{cov}\left(\varepsilon_{ct-2}, \varepsilon_{ct}\right) = \mathrm{cov}\left(\varepsilon_{ct-2}, \rho\varepsilon_{ct-1} + \upsilon_{ct}\right) = \rho^2\,\mathrm{cov}\left(\varepsilon_{ct-2}, \varepsilon_{ct-2}\right) = \frac{\sigma^2\rho^2}{\left(1 - \rho^2\right)}$$

$$\mathrm{cov}\left(\varepsilon_{ct-k}, \varepsilon_{ct}\right) = \rho^k\,\mathrm{cov}\left(\varepsilon_{ct-k}, \varepsilon_{ct-k}\right) = \frac{\sigma^2\rho^k}{\left(1 - \rho^2\right)}.$$

Inserting these expressions into the expression here and simplifying gives the expression in the text.

[5]   Another form of updating assumes that the bonus setter must make up for reenlistment shortfalls or offset excess reenlistment realized in the previous period. A reenlistment deficit in the previous period would be added to the target in the current period, resulting in a current-period bonus that is higher than it would be with a constant target. With positive

Targets with "Noise"

The bonus setter may have fresh information that affects the reenlistment target in an occupation but is unrelated to previous reenlistment outcomes. If these shocks to the target are uncorrelated with the error, they serve to increase the opportunity for estimating an unbiased bonus effect. The "noisy" target case is a simple extension of the previously discussed models. In the first model, the constant target in an occupation is replaced by $\bar{r}_{ct} = \bar{r}_c + \eta_{ct}$, and the bonus is determined to be

$$B_{ct} = \frac{\left(\bar{r}_c + \eta_{ct} - \left(\beta_0 + \mu_c + \rho\varepsilon_{ct-1}\right)\right)}{\beta_1}.$$

As before, if autocorrelation is present, the bonus variable will be correlated with the error and the bonus effect will be biased. However, the target shock increases the exogenous variation in the bonus and should increase the potential for estimating the bonus effect. In the best of worlds, autocorrelation is absent and exogenous bonus variation is ample. In the second model, an error can be added to the updating equation as follows, and the analysis can proceed as before:

$$\bar{r}_{ct} = r_{ct-1} + \gamma\left(\bar{r}_c - r_{ct-1}\right) + \eta_{ct} = \gamma\bar{r}_c + \left(1-\gamma\right)r_{ct-1} + \eta_{ct}.$$

Bonus Setting with a Target Reenlistment Number

The preceding discussion modeled the bonus setter's behavior in terms of achieving a target reenlistment rate, but the actual number of members reenlisting could be relevant. The policy goal may be to keep the number of reenlistments steady at level $\bar{n}_c$. This could be true if there were no changes in the desired size of the force and if personnel at different points in their careers (e.g., first term or second term) were not substitutes. In this case, to maintain the number of people in the force who have reached a given reenlistment point, the desired number of reenlistments would remain constant over time. Suppose the target number equals the long-run target plus a fraction of the difference between this target and last period's reenlistments:

$$\bar{n}_{ct} = \bar{n}_c + \lambda\left(\bar{n}_c - n_{ct-1}\right).$$

If $0 < \lambda < 1$, the current target level makes up part of the discrepancy between last period's reenlistments and $\bar{n}_c$.

Given m_{ct}, the number of service members coming up for reenlistment in period t, the target reenlistment rate is related to the target number of reenlistments by

$$\bar{r}_{ct} \equiv \frac{\bar{n}_{ct}}{m_{ct}} = \frac{\bar{n}_c}{m_{ct}} + \lambda\left(\frac{\bar{n}_c}{m_{ct}} - \frac{n_{ct-1}}{m_{ct}}\right).$$

---

autocorrelation, there would be a *stronger* negative association between the bonus and the error term than there would be with a constant target.

If we assume that each occupation has a long-run average number of members coming up for reenlistment, $\bar{m}_c$, then we can rewrite this expression as follows:

$$\bar{r}_{ct} = \frac{\bar{m}_c}{m_{ct}}\frac{\bar{n}_c}{\bar{m}_c} + \lambda\left(\frac{\bar{m}_c}{m_{ct}}\frac{\bar{n}_c}{\bar{m}_c} - \frac{n_{ct-1}}{m_{ct}}\right)$$

$$= (1+\lambda)\frac{\bar{m}_c}{m_{ct}}\bar{r}_c - \lambda\frac{m_{ct-1}}{m_{ct}}r_{ct-1}.$$

Plugging this expression into the bonus-setting rule gives

$$\beta_1 B_{ct} = (1+\lambda)\frac{\bar{m}_c}{m_{ct}}\bar{r}_c - \lambda\frac{m_{ct-1}}{m_{ct}}r_{ct-1} - \left(\beta_0 + \mu_c + \rho\varepsilon_{ct-1}\right)$$

$$= \left((1+\lambda)\frac{\bar{m}_c}{m_{ct}}\bar{r}_c - \left(1+\lambda\frac{m_{ct-1}}{m_{ct}}\right)(\beta_0 + \mu_c)\right) - \lambda\frac{m_{ct-1}}{m_{ct}}\beta_1 B_{ct-1} - \left(\lambda\frac{m_{ct-1}}{m_{ct}} + \rho\right)\varepsilon_{ct-1}.$$

Two insights emerge from this analysis. First, as before, if the random terms are auto-correlated, the bonus will be correlated with the contemporaneous random term through its dependence on last period's bonus and random term. This would imply that the cross-sectional relationship between the bonus and reenlistment rate would be biased. Second, all else equal, the bonus will be negatively related to the size of the cohort coming up for reenlistment. The intuition behind this result is straightforward: A smaller bonus is needed to meet the target when the pool is larger.

The latter point suggests that variation in an occupation's cohort size might be used to identify the bonus effect. Cohort size depends on recruiting targets and realizations several years earlier that may be unrelated to the current reenlistment environment. This reasoning suggests that cohort size could be an instrumental variable for bonuses, however, some experimentation with our data indicated that there was not enough variation in cohort size for this assessment to be fruitful. Moreover, a large cohort might be indicative of relatively generous recruitment incentives. With heterogeneous preferences for the military, these incentives would have induced applicants with lower taste for the military to join, and these individuals would tend to have lower propensities to reenlist. Thus, even with a relatively large cohort, reenlistment bonuses might still need to be set at a high level to maintain desired retention levels.

Omitted Variable Bias

Autocorrelation is not the only factor to keep in mind. For example, the Navy has been down-sizing and might have wanted to reduce the size of some ratings while maintaining the size of others. Bonuses would be reduced or removed in the decreasing ratings, and career counselors might reduce their effort to encourage sailors to reenlist in these ratings and increase their effort in the nondecreasing ratings. Also, the Navy might limit the slots available for reenlistment in certain occupations, possibly creating a demand constraint. These unobserved actions would be correlated with bonus and would bias up the estimate of the bonus effect. Similarly, the Marine Corps has been growing, and career counselors might have exerted more effort to retain marines in specialties offering a bonus and less effort in specialties that did not. This, too, would bias up the bonus effect. Further, marine reenlistment is sometimes described as demand-constrained, but explicit evidence of this is lacking. Relaxing the demand constraints to allow higher reenlistment in selected specialties at the same time that bonuses were increased would also bias up the bonus effect. These factors may help to explain the high first-term bonus-effect estimates that we obtained for the Navy and the Marine Corps. The Army has also been growing, but its reenlistment is not known to be demand-constrained, and Army counselors may have sustained their effort across all specialties rather than reallocating toward specialties offering bonuses. If so, Army growth would not have led to an upward bias in the bonus effect, in contrast to our conjecture about the Marine Corps.

Summary of Bonus Setting at the Occupation Level

The analysis provides several lessons for estimating bonus effects. First, because bonuses may be set, in part, to compensate for persistent features that make an MOS relatively less attractive, the reenlistment regression should include a fixed effect for each occupational specialty. Omitting the fixed effect could bias the estimate downward and even result in a negative coefficient. For instance, if bonuses were higher for less attractive occupations and the bonus did not fully compensate for this, there would be a negative correlation between bonus and reenlistment. Second, it is important to realize that bonus amounts are the result of a bonus-setting process. The models presented here are guesses at what the bonus-setting process might be, and they reveal that the bias on the bonus coefficient depends on whether the error is autocorrelated and whether the reenlistment target is updated. Assuming that fixed effects are included in the reenlistment regression, if positive autocorrelation is present and the reenlistment target is constant, the bonus coefficient will be biased *downward*. This is probably the usual case that analysts have in mind. Under the same conditions but with an updated reenlistment target, the downward bias will be smaller and could conceivably become an upward bias. The downward bias will be smaller the greater the weight placed on the reenlistment outcome in period $t-1$ when setting the reenlistment target for period t. Third, the best conditions for estimating the bonus effect are when the bonus is characterized by exogenous variation and autocorrelation is absent or small.

Finally, there could also be an omitted-variable bias resulting from the reallocation of career counselor effort—an unobserved variable—toward occupations in which bonuses increased and away from occupations in which they decreased. This would lead to an upward bias in the estimate of the bonus effect.

Reenlistment at the Individual Level

We now present a micro model of reenlistment behavior that motivates the assumption that the bonus-setting behavior outlined in the previous section is optimal and that also forms a basis for our empirical analysis.

We assume that each member has a latent propensity to reenlist, R_{ict}, that depends on the bonus, occupation fixed effect, occupation-by-period effect, and individual random term:

$$R_{ict} = \beta B_{ct} + \mu_c + \alpha_{ct} + \upsilon_{ict},$$

where υ_{ict} is the idiosyncratic random term. The individual has no influence on the bonus, but, as discussed earlier, the collective behavior of service members and expectations about aggregate reenlistment rates play a crucial role in setting the bonus. We make the additional assumption that the individual-level random terms are independent of the occupation fixed effect and occupation-by-time effect.

A member reenlists if $R_{ict} > 0$. We assume that the bonus setter knows the behavioral model, the bonus effect β, and the occupation fixed effect μ_c. Therefore, the bonus setter can set the bonus so that the marginal individual enlists:

$$\Pr\left(R_{ict} > 0\right) = \Pr\left(\alpha_{ct} + \upsilon_{ict} > -\left(\beta B_{ct} + \mu_c\right)\right).$$

The right-hand side can be expressed as

$$1 - \int_{\upsilon} \int_{-\infty}^{-\left(\beta B_{ct} + \mu_c + \upsilon_{ict}\right)} f\left(\alpha_{ct}, \upsilon_{ict}\right) d\alpha_{ct} d\upsilon_{ict},$$

where $f\left(\alpha_{ct}, \upsilon_{ict}\right)$ is the joint distribution function of the occupation-by-period and individual random terms. Since these random terms are independent of each other, we have that

$$\Pr\left(\alpha_{ct} + \upsilon_{ict} > -\left(\beta B_{ct} + \mu_c\right)\right) = 1 - \int_{\upsilon} \int_{-\infty}^{-\left(\beta B_{ct} + \mu_c + \upsilon_{ict}\right)} f\left(\alpha_{ct}, \upsilon_{ict}\right) d\alpha_{ct} d\upsilon_{ict}$$

$$= 1 - \int_{\upsilon} h\left(\upsilon_{ict}\right) G\left(-\left(\beta B_{ct} + \mu_c + \upsilon_{ict}\right)\right) d\upsilon_{ict},$$

where g and h are the marginal distribution functions of α_{ct} and υ_{ict}, respectively, and G is the cumulative distribution function of α_{ct}.

Further simplifications can be made through assumptions about the distribution of α_{ct}. In particular, suppose α_{ct} is uniformly distributed from $\bar{\alpha}_{ct} - \alpha$ to $\bar{\alpha}_{ct} + \alpha$, where $\bar{\alpha}_{ct}$ is the mean. In that case, the cumulative distribution of α_{ct} for any value x is $\left(x - \bar{\alpha}_{ct} + \alpha\right) / 2\alpha$, so the expression for the reenlistment probability simplifies to

$$1 - \left[\frac{1}{2\alpha}\left(-\left(\beta B_{ct} + \mu_c\right) - \bar{\alpha}_{ct} + \alpha\right) \int_{\upsilon} h\left(\upsilon_{ict}\right) d\upsilon_{ict} - \frac{1}{2\alpha} \int_{\upsilon} h\left(\upsilon_{ict}\right) \upsilon_{ict} d\upsilon_{ict}\right]$$

$$= 1 - \left[\frac{1}{2\alpha}\left(-\left(\beta B_{ct} + \mu_c\right) - \bar{\alpha}_{ct} + \alpha\right) - E\left(\upsilon_{ict}\right)\right].$$

Relating this framework to the occupation-level model, if the occupation-by-period shocks are autocorrelated, then $\bar{\alpha}_{ct} = \rho \alpha_{ct-1}$. Essentially, we are assuming that the new shock in the current period, α_{ct}, is uniformly distributed with a mean of zero; in the current period, the last period's shock is observed and nonrandom. The assumption of a zero mean is justifiable, given that the model includes a fixed effect for occupation, μ_c. Further, we assume $E\left(\upsilon_{ict}\right) = 0$, so it vanishes from the previous expression. We then have

$$\Pr\left(\alpha_{ct} + \upsilon_{ict} > -\left(\beta B_{ct} + \mu_{c}\right)\right) = \frac{1}{2} - \frac{1}{2\alpha}\left(-\left(\beta B_{ct} + \mu_{c}\right) - \rho\alpha_{ct-1}\right).$$

If α is normalized to be equal to 0.5, this expression is the same as the expected reenlistment rate given in the previous section, with $\beta_{0} = 1/2$. That is,

$$\Pr\left(\alpha_{ct} + \upsilon_{ict} > -\left(\beta B_{ct} + \mu_{c}\right)\right) = \frac{1}{2} + \beta B_{ct} + \mu_{c} + \rho\alpha_{ct-1}.$$

Also note that, even without making any assumption about the distribution of the occupation-by-period shocks, it can be shown using the implicit function theorem that the optimal bonus is increasing in the desired retention rate.

Relationship Between Bias in Estimated Bonus Effect and Estimated Deployment Effect

This appendix presents the mathematical formulation of the potential biases of the estimated bonus and deployment coefficients. To begin, note that the Frisch-Waugh-Lovell theorem implies that least squares estimates of the basic estimation equation (5.1 in Chapter Five) will be equal to least squares estimates of the model:

$$R_i^* = \theta D_i^* + \beta B_i^* + \varepsilon_i^*,$$

where R_i^* is the residual of a least squares regression of R_i on X_i, and the other variables are defined in the same way. As argued in Chapter Four, bonuses are likely to be correlated with unobservable determinants of reenlistment, even after controlling for covariates, including occupation fixed effects. Such correlations would render estimates of β inconsistent.

To see whether estimates of the effect of deployments will also be inconsistent, note that the probability limit of the regression estimate of θ is equal to

$$\theta + \frac{\sigma_{B^*}^2 \sigma_{D^*\varepsilon} - \sigma_{B^*\varepsilon}\sigma_{D^*B^*}}{\sigma_{B^*}^2 \sigma_{D^*}^2 \left(1 - \rho_{D^*B^*}^2\right)},$$

where $\sigma_{D^*}^2$ and $\sigma_{B^*}^2$ are the variances of D^* and B^*, respectively; $\sigma_{D^*B^*}$ and $\sigma_{B^*\varepsilon^*}$ are the covariances between D^* and B^* and between ε^* and B^*, respectively; and $\rho_{D^*B^*}$ is the correlation between D^* and B^*. Assuming that deployments are exogenous to the individual (after controlling for X_i, which includes MOS fixed effects) and thus that $\sigma_{D^*\varepsilon^*} = 0$, the inconsistency of θ is determined by the term $-\sigma_{D^*B^*}\sigma_{B^*\varepsilon^*}$. If deployments and reenlistment bonuses are correlated (after controlling for X_i), then the regression estimate of θ will be inconsistent if bonuses are endogenous (i.e., when $\sigma_{B^*\varepsilon^*} \neq 0$). In particular, if bonus and deployments are positively related and bonuses are negatively related to ε^*, then $-\sigma_{D^*B^*}\sigma_{B^*\varepsilon^*} > 0$ and the estimates of θ will be overstated.

Additional Regression Results

This appendix presents the regression results from the models used in our analysis. Table C.1 presents a glossary of the variables used in the regression analysis. Tables C.2 through C.9 show the effect of deployment on work and personal stress, intention to reenlist, and reenlistment among first- and second-term-plus survey respondents. Tables C.10 through C.17 show the same results with controls for overtime and time away that was more or less than expected. Tables C.18 through C.29 show the effect of deployment on reenlistment without MOS controls and with MOS fixed effects and MOS-by-quarter fixed effects in the administrative data for one-year and three-year windows.

Tables C.30 through C.42 break out the results by year of reenlistment decision and by service. Tables C.43 through C.48 show the effects of deployment on reenlistment by gender or occupation and year of decision for each service. Tables C.49 through C.51 present descriptive statistics for the administrative and survey data samples.

Table C.1
Glossary of Variables

Variable	Description
Nonhostile deployment only in window	1 if deployed in 12- or 36-month window but did not receive HFP, 0 otherwise
Hostile deployment in window	1 if received HFP in 12-or 36-month window
SRB multiplier	Average bonus multiplier in respondent's occupation and zone at time of reenlistment decision
Years of service	Years of active-duty military service
HS dropout or education missing	1 if respondent has less than a HS diploma or education information is missing, 0 otherwise
GED	1 if respondent has a General Educational Development (GED) degree, 0 otherwise
At least some college	1 if respondent has at least some college education, 0 otherwise
Male	1 if respondent is male, 0 otherwise
AFQT less than category IIIB	1 if AFQT percentile below 31st percentile
AFQT category IIIB	1 if AFQT percentile between 31st and 49th percentile
AFQT category II	1 if AFQT percentile between 65th and 92nd percentile
AFQT category I	1 if AFQT percentile above 92nd percentile
White, black, Hispanic, or other	1 if respondent is member of given racial group
Promoted rapidly	1 if pay grade is higher than the average pay grade for members with same years of service at time of reenlistment decision
Away from home but not deployed	1 if respondent spent any nights away from usual duty location in 12 months prior to survey, but not deployed, 0 otherwise
Very poorly prepared for job duties	1 if respondent indicated that he or she was very poorly prepared for his or her wartime job duty, 0 otherwise
Poorly prepared for job duties	1 if respondent indicated that he or she was poorly prepared for his or her wartime job duty, 0 otherwise
Well prepared for job duties	1 if respondent indicated that he or she was well prepared for his or her wartime job duty, 0 otherwise
Very well prepared for job duties	1 if respondent indicated that he or she was very well prepared for his or her wartime job duty, 0 otherwise

Table C.2
Effect of Deployment on Work Stress, First-Term Survey Respondents

Variable	Army	Navy	Marine Corps	Air Force
Nonhostile deployment only	−0.019	0.018	0.015	−0.035
	(0.023)	(0.026)	(0.028)	(0.030)
Hostile deployment	0.115**	0.075**	0.044*	−0.018
	(0.017)	(0.016)	(0.019)	(0.018)
Away from home but not deployed	0.048**	0.063**	0.045**	0.041**
	(0.015)	(0.015)	(0.016)	(0.016)
Very poorly prepared for job	0.115**	0.203**	0.165**	0.175**
	(0.032)	(0.040)	(0.049)	(0.059)
Poorly prepared for job	0.053*	0.098**	0.129**	0.023
	(0.025)	(0.033)	(0.036)	(0.037)
Well prepared for job	−0.037*	−0.080**	0.006	−0.041*
	(0.018)	(0.020)	(0.022)	(0.021)
Very well prepared for job	−0.026	−0.062**	0.056*	−0.046*
	(0.020)	(0.022)	(0.024)	(0.023)
AFQT category I	−0.010	0.001	−0.024	−0.043
	(0.040)	(0.050)	(0.058)	(0.049)
AFQT category II	0.027	0.008	−0.026	0.018
	(0.035)	(0.041)	(0.051)	(0.040)
AFQT category IIIA	0.027	0.013	−0.001	0.022
	(0.035)	(0.042)	(0.052)	(0.041)
AFQT category IIIB	0.013	−0.000	−0.023	0.018
	(0.036)	(0.042)	(0.052)	(0.042)
Some college	−0.006	−0.046	−0.066	−0.011
	(0.036)	(0.054)	(0.061)	(0.023)
College graduate	−0.056*	−0.056	−0.057	−0.043
	(0.024)	(0.041)	(0.069)	(0.055)
Black	−0.041*	−0.037†	−0.017	−0.021
	(0.017)	(0.020)	(0.024)	(0.019)
Hispanic	−0.058**	−0.044*	−0.065**	−0.066*
	(0.018)	(0.021)	(0.020)	(0.027)
Other race	0.002	−0.048*	0.019	−0.046†
	(0.024)	(0.020)	(0.030)	(0.025)
Married	0.022	0.010	−0.006	0.046**
	(0.015)	(0.018)	(0.019)	(0.017)
Number of observations	6,294	5,160	4,688	5,250

NOTE: Models also include controls for location (rural/urban), dual-service career spouse, gender, pay grade, years of service, survey wave indicators, and one-digit DoD occupation. ** = statistically significant at the 1-percent level. * = statistically significant at the 5-percent level. † = statistically significant at the 10-percent level.

Table C.3
Effect of Deployment on Work Stress, Second-Term-Plus Survey Respondents

Variable	Army	Navy	Marine Corps	Air Force
Nonhostile deployment only	−0.000	0.075**	−0.024	0.013
	(0.014)	(0.016)	(0.019)	(0.017)
Hostile deployment	0.098**	0.126**	0.054**	0.020
	(0.012)	(0.013)	(0.019)	(0.014)
Away from home but not deployed	0.021*	0.075**	0.063**	0.026*
	(0.011)	(0.011)	(0.015)	(0.012)
Very poorly prepared for job	0.147**	0.113*	0.227**	0.221**
	(0.037)	(0.054)	(0.069)	(0.052)
Poorly prepared for job	0.087**	0.085**	0.060	0.060*
	(0.025)	(0.032)	(0.044)	(0.029)
Well prepared for job	−0.082**	−0.057**	−0.032	−0.034†
	(0.015)	(0.018)	(0.024)	(0.017)
Very well prepared for job	−0.089**	−0.063**	−0.035	−0.043*
	(0.016)	(0.018)	(0.024)	(0.018)
AFQT category I	0.035	0.059†	0.048	−0.086*
	(0.031)	(0.031)	(0.048)	(0.040)
AFQT category II	0.051**	0.006	0.110**	−0.060†
	(0.019)	(0.021)	(0.035)	(0.032)
AFQT category IIIA	0.038*	−0.018	0.099**	−0.083*
	(0.019)	(0.021)	(0.035)	(0.032)
AFQT category IIIB	0.025	−0.048*	0.058†	−0.076*
	(0.018)	(0.021)	(0.035)	(0.033)
Some college	−0.006	−0.010	−0.012	0.062**
	(0.014)	(0.027)	(0.031)	(0.023)
College graduate	−0.017	−0.008	−0.011	0.029
	(0.018)	(0.024)	(0.035)	(0.031)
Black	−0.078**	−0.053**	−0.075**	−0.066**
	(0.011)	(0.014)	(0.017)	(0.013)
Hispanic	−0.020	−0.031†	−0.050*	−0.042†
	(0.015)	(0.016)	(0.020)	(0.022)
Other race	−0.046*	−0.055**	−0.034	−0.011
	(0.019)	(0.016)	(0.030)	(0.021)
Married	0.014	0.028*	−0.011	−0.001
	(0.012)	(0.012)	(0.018)	(0.013)
Number of observations	12,578	9,771	5,673	8,845

NOTE: Models also include controls for location (rural/urban), dual-service career spouse, gender, pay grade, years of service, survey wave indicators, and one-digit DoD occupation. ** = statistically significant at the 1-percent level. * = statistically significant at the 5-percent level. † = statistically significant at the 10-percent level.

Table C.4
Effect of Deployment on Personal Stress, First-Term Survey Respondents

Variable	Army	Navy	Marine Corps	Air Force
Nonhostile deployment only	0.035	0.047[†]	0.010	0.039
	(0.023)	(0.027)	(0.028)	(0.030)
Hostile deployment	0.073**	0.065**	0.026	0.022
	(0.017)	(0.017)	(0.020)	(0.018)
Away from home but not deployed	0.015	0.026[†]	0.002	0.027[†]
	(0.015)	(0.015)	(0.017)	(0.015)
Very poorly prepared for job	0.106**	0.079	0.085	0.107[†]
	(0.034)	(0.051)	(0.052)	(0.063)
Poorly prepared for job	0.041	0.084*	0.111**	0.123**
	(0.026)	(0.035)	(0.036)	(0.037)
Well prepared for job	−0.026	−0.008	−0.040[†]	−0.004
	(0.018)	(0.020)	(0.022)	(0.020)
Very well prepared for job	−0.026	−0.015	0.003	−0.006
	(0.020)	(0.022)	(0.024)	(0.022)
AFQT category I	−0.019	0.007	−0.015	0.028
	(0.041)	(0.050)	(0.057)	(0.048)
AFQT category II	0.022	0.031	0.048	0.012
	(0.036)	(0.042)	(0.051)	(0.039)
AFQT category IIIA	0.009	0.009	0.036	−0.010
	(0.036)	(0.042)	(0.051)	(0.040)
AFQT category IIIB	−0.006	0.000	0.031	−0.039
	(0.037)	(0.042)	(0.052)	(0.040)
Some college	−0.056	−0.124*	−0.076	0.004
	(0.036)	(0.050)	(0.061)	(0.022)
College graduate	−0.051*	−0.066[†]	0.070	0.018
	(0.024)	(0.040)	(0.068)	(0.054)
Black	−0.010	−0.054**	0.019	−0.015
	(0.018)	(0.020)	(0.024)	(0.018)
Hispanic	0.018	0.023	0.017	−0.001
	(0.018)	(0.022)	(0.020)	(0.027)
Other race	0.044[†]	0.020	−0.052[†]	−0.036
	(0.024)	(0.021)	(0.031)	(0.024)
Married	0.061**	0.058**	0.016	0.028[†]
	(0.015)	(0.018)	(0.019)	(0.016)
Number of observations	6,294	5,160	4,688	5,250

NOTE: Models also include controls for location (rural/urban), dual-service career spouse, gender, pay grade, years of service, survey wave indicators, and one-digit DoD occupation. ** = statistically significant at the 1-percent level. * = statistically significant at the 5-percent level. † = statistically significant at the 10-percent level.

Table C.5
Effect of Deployment on Personal Stress, Second-Term-Plus Survey Respondents

Variable	Army	Navy	Marine Corps	Air Force
Nonhostile deployment only	0.056**	0.048**	0.020	0.050**
	(0.014)	(0.016)	(0.019)	(0.017)
Hostile deployment	0.093**	0.038**	0.037*	0.029*
	(0.012)	(0.013)	(0.018)	(0.014)
Away from home but not deployed	0.004	0.024*	0.020	0.025*
	(0.011)	(0.010)	(0.015)	(0.011)
Very poorly prepared for job	0.068†	0.135*	0.194**	0.145*
	(0.040)	(0.056)	(0.070)	(0.057)
Poorly prepared for job	0.051*	0.053	0.032	0.030
	(0.026)	(0.033)	(0.045)	(0.029)
Well prepared for job	−0.086**	−0.038*	−0.043†	−0.033*
	(0.015)	(0.018)	(0.024)	(0.017)
Very well prepared for job	−0.104**	−0.064**	−0.082**	−0.058**
	(0.016)	(0.018)	(0.024)	(0.018)
AFQT category I	0.066*	0.027	0.085†	−0.026
	(0.030)	(0.030)	(0.048)	(0.040)
AFQT category II	0.074**	0.013	0.091**	−0.059†
	(0.019)	(0.021)	(0.035)	(0.032)
AFQT category IIIA	0.051**	−0.007	0.076*	−0.073*
	(0.019)	(0.021)	(0.035)	(0.032)
AFQT category IIIB	0.040*	−0.032	0.056	−0.067*
	(0.018)	(0.020)	(0.034)	(0.032)
Some college	−0.016	−0.025	−0.016	0.040†
	(0.014)	(0.026)	(0.031)	(0.022)
College graduate	−0.036*	−0.016	0.002	0.029
	(0.017)	(0.023)	(0.034)	(0.029)
Black	−0.060**	−0.051**	−0.038*	−0.070**
	(0.011)	(0.013)	(0.017)	(0.013)
Hispanic	−0.006	−0.001	−0.036†	0.019
	(0.015)	(0.016)	(0.019)	(0.022)
Other race	−0.015	−0.028†	0.014	−0.001
	(0.019)	(0.016)	(0.030)	(0.021)
Married	0.047**	0.038**	0.045**	0.010
	(0.011)	(0.012)	(0.017)	(0.013)
Number of observations	12,578	9,771	5,673	8,845

NOTE: Models also include controls for location (rural/urban), dual-service career spouse, gender, pay grade, years of service, survey wave indicators, and one-digit DoD occupation. ** = statistically significant at the 1-percent level. * = statistically significant at the 5-percent level. † = statistically significant at the 10-percent level.

Table C.6
Effect of Deployment on Intention to Reenlist, First-Term Survey Respondents

Variable	Army	Navy	Marine Corps	Air Force
Nonhostile deployment only	0.050*	0.027	0.005	0.024
	(0.021)	(0.025)	(0.026)	(0.028)
Hostile deployment	−0.102**	−0.044**	−0.095**	−0.062**
	(0.015)	(0.016)	(0.018)	(0.018)
Away from home but not deployed	−0.004	−0.018	−0.012	−0.003
	(0.014)	(0.015)	(0.015)	(0.015)
Very poorly prepared for job	−0.080**	−0.173**	−0.125**	−0.193**
	(0.027)	(0.039)	(0.038)	(0.051)
Poorly prepared for job	−0.002	−0.045	−0.050†	−0.039
	(0.022)	(0.032)	(0.030)	(0.035)
Well prepared for job	0.130**	0.116**	0.079**	0.109**
	(0.016)	(0.019)	(0.020)	(0.020)
Very well prepared for job	0.150**	0.204**	0.135**	0.159**
	(0.018)	(0.021)	(0.021)	(0.022)
AFQT category I	−0.107**	−0.103*	−0.047	−0.052
	(0.036)	(0.049)	(0.049)	(0.048)
AFQT category II	−0.131**	−0.134**	−0.015	0.012
	(0.032)	(0.041)	(0.044)	(0.039)
AFQT category IIIA	−0.115**	−0.088*	0.022	0.004
	(0.032)	(0.041)	(0.045)	(0.040)
AFQT category IIIB	−0.075*	−0.056	0.024	0.039
	(0.033)	(0.041)	(0.046)	(0.041)
Some college	−0.017	0.087†	0.082	−0.006
	(0.032)	(0.053)	(0.058)	(0.021)
College graduate	−0.041*	0.040	−0.068	0.030
	(0.021)	(0.040)	(0.060)	(0.054)
Black	0.055**	0.062**	0.051*	0.057**
	(0.016)	(0.020)	(0.022)	(0.018)
Hispanic	0.021	0.058**	0.052**	−0.013
	(0.016)	(0.021)	(0.019)	(0.026)
Other race	−0.016	0.047*	0.044	0.049*
	(0.021)	(0.020)	(0.028)	(0.024)
Married	0.070**	0.092**	0.076**	0.083**
	(0.014)	(0.018)	(0.017)	(0.016)
Number of observations	6,720	5,339	4,927	5,554

NOTE: Models also include controls for location (rural/urban), dual-service career spouse, gender, pay grade, years of service, survey wave indicators, and one-digit DoD occupation. ** = statistically significant at the 1-percent level. * = statistically significant at the 5-percent level. † = statistically significant at the 10-percent level.

Table C.7
Effect of Deployment on Intention to Reenlist, Second-Term-Plus Survey Respondents

Variable	Army	Navy	Marine Corps	Air Force
Nonhostile deployment only	−0.003	0.014	0.016	−0.001
	(0.012)	(0.014)	(0.016)	(0.014)
Hostile deployment	−0.051**	0.021†	−0.009	−0.036**
	(0.011)	(0.011)	(0.016)	(0.012)
Away from home but not deployed	−0.002	−0.008	0.009	0.004
	(0.009)	(0.009)	(0.013)	(0.009)
Very poorly prepared for job	−0.198**	−0.150**	−0.286**	−0.256**
	(0.036)	(0.051)	(0.070)	(0.052)
Poorly prepared for job	−0.038	−0.079*	−0.026	−0.053*
	(0.025)	(0.031)	(0.040)	(0.027)
Well prepared for job	0.112**	0.117**	0.078**	0.092**
	(0.014)	(0.016)	(0.022)	(0.015)
Very well prepared for job	0.154**	0.134**	0.101**	0.114**
	(0.015)	(0.017)	(0.022)	(0.016)
AFQT category I	−0.028	−0.130**	0.002	−0.040
	(0.028)	(0.027)	(0.044)	(0.033)
AFQT category II	−0.016	−0.063**	−0.007	−0.013
	(0.017)	(0.017)	(0.032)	(0.026)
AFQT category IIIA	0.009	−0.006	0.001	−0.017
	(0.017)	(0.017)	(0.032)	(0.026)
AFQT category IIIB	−0.009	0.008	0.010	−0.033
	(0.016)	(0.016)	(0.031)	(0.026)
Some college	0.016	0.007	−0.005	−0.072**
	(0.013)	(0.023)	(0.026)	(0.020)
College graduate	−0.023	−0.013	−0.130**	−0.110**
	(0.016)	(0.020)	(0.033)	(0.026)
Black	0.002	0.003	0.006	−0.005
	(0.010)	(0.012)	(0.015)	(0.011)
Hispanic	0.001	0.014	0.018	−0.012
	(0.014)	(0.014)	(0.017)	(0.019)
Other race	0.014	0.025†	−0.032	0.022
	(0.017)	(0.014)	(0.026)	(0.017)
Married	0.032**	0.042**	0.038*	0.033**
	(0.011)	(0.011)	(0.016)	(0.012)
Number of observations	13,898	10,452	6,094	9,948

NOTE: Models also include controls for location (rural/urban), dual-service career spouse, gender, pay grade, years of service, survey wave indicators, and one-digit DoD occupation. ** = statistically significant at the 1-percent level. * = statistically significant at the 5-percent level. † = statistically significant at the 10-percent level.

Table C.8
Effect of Deployment on Reenlistment, First-Term Survey Respondents

Variable	Army	Navy	Marine Corps	Air Force
Nonhostile deployment only	−0.002	0.017	0.014	0.114**
	(0.031)	(0.033)	(0.028)	(0.035)
Hostile deployment	−0.056**	−0.012	−0.031	0.025
	(0.020)	(0.020)	(0.019)	(0.022)
Away from home but not deployed	0.003	0.011	−0.004	−0.012
	(0.019)	(0.019)	(0.017)	(0.019)
Very poorly prepared for job	−0.062	−0.225**	−0.124**	−0.125[†]
	(0.041)	(0.052)	(0.041)	(0.066)
Poorly prepared for job	−0.014	−0.079[†]	−0.004	0.053
	(0.032)	(0.045)	(0.035)	(0.044)
Well prepared for job	0.019	0.053*	0.038[†]	0.071**
	(0.022)	(0.025)	(0.022)	(0.025)
Very well prepared for job	0.051*	0.126**	0.043[†]	0.073**
	(0.024)	(0.027)	(0.023)	(0.027)
AFQT category I	−0.059	−0.060	−0.031	−0.211**
	(0.051)	(0.070)	(0.061)	(0.061)
AFQT category II	−0.091*	−0.086	−0.008	−0.156**
	(0.043)	(0.054)	(0.055)	(0.049)
AFQT category IIIA	−0.059	−0.007	0.033	−0.128**
	(0.043)	(0.054)	(0.056)	(0.049)
AFQT category IIIB	−0.051	0.041	0.040	−0.064
	(0.044)	(0.053)	(0.056)	(0.051)
Some college	−0.077[†]	0.144**	0.011	0.000
	(0.042)	(0.055)	(0.060)	(0.024)
College graduate	−0.117**	−0.093	−0.128*	0.162*
	(0.030)	(0.060)	(0.055)	(0.066)
Black	0.111**	0.132**	0.125**	0.125**
	(0.021)	(0.024)	(0.025)	(0.022)
Hispanic	0.032	0.080**	0.026	0.081**
	(0.022)	(0.025)	(0.020)	(0.031)
Other race	0.046	0.094**	0.056[†]	0.072*
	(0.029)	(0.025)	(0.030)	(0.030)
Married	0.116**	0.110**	0.106**	0.102**
	(0.019)	(0.022)	(0.018)	(0.021)
Number of observations	3,954	3,228	3,728	3,516

NOTE: Models also include controls for location (rural/urban), dual-service career spouse, gender, pay grade, years of service, survey wave indicators, and one-digit DoD occupation. ** = statistically significant at the 1-percent level. * = statistically significant at the 5-percent level. [†] = statistically significant at the 10-percent level.

Table C.9
Effect of Deployment on Reenlistment, Second-Term-Plus Survey Respondents

Variable	Army	Navy	Marine Corps	Air Force
Nonhostile deployment only	0.028	0.107**	0.105**	0.037†
	(0.020)	(0.022)	(0.021)	(0.019)
Hostile deployment	0.008	0.118**	0.086**	0.028†
	(0.017)	(0.017)	(0.023)	(0.017)
Away from home but not deployed	0.018	0.052**	0.023	0.026*
	(0.015)	(0.014)	(0.017)	(0.013)
Very poorly prepared for job	−0.168**	−0.104	−0.307**	−0.178*
	(0.055)	(0.068)	(0.076)	(0.079)
Poorly prepared for job	−0.074*	−0.043	−0.020	−0.018
	(0.036)	(0.042)	(0.052)	(0.037)
Well prepared for job	0.049*	0.048*	0.056†	0.056**
	(0.022)	(0.023)	(0.029)	(0.021)
Very well prepared for job	0.059**	0.034	0.058†	0.031
	(0.022)	(0.023)	(0.030)	(0.022)
AFQT category I	−0.009	0.015	0.083	0.154**
	(0.046)	(0.040)	(0.058)	(0.048)
AFQT category II	0.012	0.051*	0.095*	0.179**
	(0.025)	(0.026)	(0.043)	(0.038)
AFQT category IIIA	0.084**	0.078**	0.093*	0.176**
	(0.024)	(0.026)	(0.043)	(0.038)
AFQT category IIIB	0.077**	0.077**	0.083*	0.158**
	(0.023)	(0.025)	(0.042)	(0.038)
Some college	−0.013	−0.054	−0.028	−0.025
	(0.023)	(0.033)	(0.037)	(0.048)
College graduate	−0.052†	−0.011	−0.114**	−0.040
	(0.028)	(0.030)	(0.038)	(0.054)
Black	0.049**	0.035*	0.037†	0.033*
	(0.016)	(0.017)	(0.020)	(0.015)
Hispanic	0.014	0.062**	0.049*	0.009
	(0.021)	(0.021)	(0.023)	(0.026)
Other race	0.019	0.070**	0.020	0.017
	(0.026)	(0.021)	(0.034)	(0.026)
Married	0.081**	0.040*	0.036†	−0.005
	(0.018)	(0.016)	(0.022)	(0.017)
Number of observations	5,208	5,663	3,809	5,176

NOTE: Models also include controls for location (rural/urban), dual-service career spouse, gender, pay grade, years of service, survey wave indicators, and one-digit DoD occupation. ** = statistically significant at the 1-percent level. * = statistically significant at the 5-percent level. † = statistically significant at the 10-percent level.

Table C.10
Effect of Deployment on Work Stress, First-Term Survey Respondents, with Controls for Overtime and Time Away More or Less than Expected

Variable	Army	Navy	Marine Corps	Air Force
Nonhostile deployment only	−0.019	0.000	0.003	−0.035
	(0.023)	(0.025)	(0.027)	(0.029)
Hostile deployment	0.010	0.010	−0.042*	−0.063**
	(0.018)	(0.017)	(0.020)	(0.019)
Away from home but not deployed	−0.001	−0.007	−0.002	0.002
	(0.015)	(0.015)	(0.016)	(0.016)
21–60 overtime days	0.157**	0.172**	0.144**	0.140**
	(0.018)	(0.018)	(0.021)	(0.018)
61–120 overtime days	0.221**	0.233**	0.232**	0.187**
	(0.020)	(0.022)	(0.022)	(0.023)
More than 120 overtime days	0.270**	0.313**	0.286**	0.337**
	(0.017)	(0.018)	(0.020)	(0.019)
Much less time away than expected	−0.025	0.063**	−0.012	−0.007
	(0.024)	(0.025)	(0.022)	(0.019)
Less time away than expected	0.016	0.041†	0.021	0.026
	(0.022)	(0.023)	(0.022)	(0.019)
More time away than expected	0.061**	0.105**	0.043†	0.028
	(0.018)	(0.019)	(0.022)	(0.024)
Much more time away than expected	0.099**	0.120**	0.100**	0.046
	(0.019)	(0.021)	(0.025)	(0.032)
Very poorly prepared for job	0.104**	0.159**	0.136**	0.137*
	(0.031)	(0.040)	(0.047)	(0.056)
Poorly prepared for job	0.058*	0.104**	0.106**	0.007
	(0.024)	(0.032)	(0.035)	(0.037)
Well prepared for job	−0.027	−0.071**	−0.002	−0.042*
	(0.017)	(0.019)	(0.022)	(0.020)
Very well prepared for job	−0.041*	−0.076**	0.018	−0.074**
	(0.019)	(0.022)	(0.024)	(0.022)
AFQT category I	−0.025	−0.015	−0.037	−0.044
	(0.040)	(0.048)	(0.058)	(0.048)
AFQT category II	0.012	−0.014	−0.031	0.012
	(0.034)	(0.040)	(0.052)	(0.038)
AFQT category IIIA	0.018	−0.009	0.004	0.025
	(0.035)	(0.040)	(0.052)	(0.039)
AFQT category IIIB	0.023	0.004	−0.002	0.031
	(0.035)	(0.040)	(0.053)	(0.040)

Table C.10—Continued

Variable	Army	Navy	Marine Corps	Air Force
Some college	−0.016	−0.031	−0.060	−0.006
	(0.036)	(0.052)	(0.062)	(0.022)
College graduate	−0.040[†]	−0.042	−0.044	−0.050
	(0.023)	(0.040)	(0.070)	(0.052)
Black	−0.012	−0.014	−0.007	0.004
	(0.017)	(0.020)	(0.023)	(0.018)
Hispanic	−0.044*	−0.050*	−0.060**	−0.055*
	(0.018)	(0.021)	(0.020)	(0.027)
Other race	0.007	−0.038[†]	0.046	−0.031
	(0.024)	(0.020)	(0.029)	(0.024)
Married	0.017	0.003	−0.011	0.039*
	(0.015)	(0.018)	(0.019)	(0.016)
Number of observations	6,212	5,090	4,628	5,194

NOTE: Models also include controls for location (rural/urban), dual-service career spouse, gender, pay grade, years of service, survey wave indicators, and one-digit DoD occupation. ** = statistically significant at the 1-percent level. * = statistically significant at the 5-percent level. [†] = statistically significant at the 10-percent level.

Table C.11
Effect of Deployment on Work Stress, Second-Term-Plus Survey Respondents, with Controls for Overtime and Time Away More or Less than Expected

Variable	Army	Navy	Marine Corps	Air Force
Nonhostile deployment only	−0.026[†]	0.027[†]	−0.042*	−0.010
	(0.013)	(0.016)	(0.019)	(0.017)
Hostile deployment	−0.025*	0.012	−0.072**	−0.070**
	(0.012)	(0.013)	(0.020)	(0.015)
Away from home but not deployed	−0.010	0.012	0.023	0.002
	(0.010)	(0.010)	(0.015)	(0.011)
21–60 overtime days	0.143**	0.173**	0.114**	0.169**
	(0.013)	(0.013)	(0.018)	(0.014)
61–120 overtime days	0.226**	0.263**	0.200**	0.257**
	(0.014)	(0.016)	(0.021)	(0.016)
More than 120 overtime days	0.317**	0.372**	0.312**	0.328**
	(0.012)	(0.013)	(0.017)	(0.014)
Much less time away than expected	−0.034*	−0.013	−0.076**	−0.065**
	(0.017)	(0.018)	(0.021)	(0.017)
Less time away than expected	−0.044**	−0.018	−0.026	−0.032*
	(0.015)	(0.016)	(0.019)	(0.015)
More time away than expected	0.061**	0.066**	0.082**	0.096**
	(0.013)	(0.015)	(0.021)	(0.018)

Table C.11—Continued

Variable	Army	Navy	Marine Corps	Air Force
Much more time away than expected	0.143**	0.140**	0.126**	0.124**
	(0.013)	(0.018)	(0.026)	(0.023)
Very poorly prepared for job	0.105**	0.066	0.208**	0.186**
	(0.037)	(0.052)	(0.071)	(0.050)
Poorly prepared for job	0.077**	0.075*	0.050	0.043
	(0.024)	(0.031)	(0.042)	(0.028)
Well prepared for job	−0.077**	−0.045**	−0.030	−0.040*
	(0.015)	(0.017)	(0.023)	(0.017)
Very well prepared for job	−0.105**	−0.071**	−0.043†	−0.068**
	(0.015)	(0.017)	(0.024)	(0.018)
AFQT category I	0.013	0.026	0.026	−0.089*
	(0.030)	(0.029)	(0.047)	(0.039)
AFQT category II	0.027	−0.022	0.082*	−0.068*
	(0.019)	(0.020)	(0.035)	(0.031)
AFQT category IIIA	0.020	−0.038†	0.083*	−0.084**
	(0.019)	(0.021)	(0.035)	(0.031)
AFQT category IIIB	0.017	−0.049*	0.054	−0.069*
	(0.018)	(0.020)	(0.035)	(0.031)
Some college	0.001	−0.009	−0.015	0.057*
	(0.014)	(0.026)	(0.030)	(0.022)
College graduate	−0.001	0.000	0.013	0.023
	(0.017)	(0.023)	(0.034)	(0.029)
Black	−0.051**	−0.016	−0.043*	−0.035**
	(0.011)	(0.013)	(0.017)	(0.013)
Hispanic	−0.012	−0.033*	−0.039*	−0.034
	(0.014)	(0.016)	(0.019)	(0.021)
Other race	−0.029	−0.046**	−0.026	0.003
	(0.018)	(0.016)	(0.029)	(0.021)
Married	0.010	0.022†	−0.017	−0.003
	(0.011)	(0.012)	(0.017)	(0.013)
Number of observations	12,440	9,681	5,608	8,774

NOTE: Models also include controls for location (rural/urban), dual-service career spouse, gender, pay grade, years of service, survey wave indicators, and one-digit DoD occupation. ** = statistically significant at the 1-percent level. * = statistically significant at the 5-percent level. † = statistically significant at the 10-percent level.

Table C.12
Effect of Deployment on Personal Stress, First-Term Survey Respondents, with Controls for Overtime and Time Away More or Less than Expected

Variable	Army	Navy	Marine Corps	Air Force
Nonhostile deployment only	0.034	0.030	0.002	0.039
	(0.023)	(0.027)	(0.028)	(0.030)
Hostile deployment	0.017	0.024	−0.008	0.000
	(0.019)	(0.017)	(0.021)	(0.019)
Away from home but not deployed	−0.005	−0.013	−0.020	0.010
	(0.016)	(0.016)	(0.017)	(0.016)
21–60 overtime days	0.064**	0.046*	0.076**	0.067**
	(0.018)	(0.018)	(0.021)	(0.017)
61–120 overtime days	0.080**	0.114**	0.125**	0.047*
	(0.021)	(0.023)	(0.023)	(0.022)
More than 120 overtime days	0.095**	0.120**	0.117**	0.078**
	(0.018)	(0.019)	(0.020)	(0.019)
Much less time away than expected	−0.004	0.004	−0.001	−0.020
	(0.024)	(0.026)	(0.023)	(0.019)
Less time away than expected	−0.001	−0.003	0.001	0.003
	(0.022)	(0.023)	(0.022)	(0.019)
More time away than expected	0.047*	0.071**	0.019	0.000
	(0.019)	(0.021)	(0.023)	(0.024)
Much more time away than expected	0.100**	0.105**	0.025	0.084*
	(0.021)	(0.024)	(0.027)	(0.033)
Very poorly prepared for job	0.097**	0.055	0.064	0.099
	(0.034)	(0.051)	(0.052)	(0.062)
Poorly prepared for job	0.047[†]	0.086*	0.099**	0.117**
	(0.026)	(0.035)	(0.037)	(0.037)
Well prepared for job	−0.021	−0.006	−0.045*	−0.003
	(0.018)	(0.020)	(0.022)	(0.020)
Very well prepared for job	−0.032	−0.022	−0.017	−0.011
	(0.020)	(0.022)	(0.024)	(0.022)
AFQT category I	−0.013	0.010	−0.023	0.020
	(0.042)	(0.049)	(0.059)	(0.048)
AFQT category II	0.024	0.026	0.045	0.008
	(0.036)	(0.041)	(0.052)	(0.039)
AFQT category IIIA	0.014	0.003	0.033	−0.010
	(0.037)	(0.041)	(0.053)	(0.039)
AFQT category IIIB	0.005	0.007	0.036	−0.034
	(0.037)	(0.041)	(0.054)	(0.040)

Table C.12—Continued

Variable	Army	Navy	Marine Corps	Air Force
Some college	−0.072*	−0.114*	−0.075	0.007
	(0.036)	(0.049)	(0.062)	(0.022)
College graduate	−0.046[†]	−0.058	0.081	0.013
	(0.024)	(0.039)	(0.067)	(0.054)
Black	0.001	−0.049*	0.023	−0.008
	(0.018)	(0.020)	(0.024)	(0.018)
Hispanic	0.023	0.017	0.018	0.006
	(0.018)	(0.022)	(0.021)	(0.027)
Other race	0.052*	0.023	−0.048	−0.033
	(0.024)	(0.021)	(0.031)	(0.024)
Married	0.058**	0.056**	0.015	0.026
	(0.015)	(0.018)	(0.019)	(0.016)
Number of observations	6,212	5,090	4,628	5,194

NOTE: Models also include controls for location (rural/urban), dual-service career spouse, gender, pay grade, years of service, survey wave indicators, and one-digit DoD occupation. ** = statistically significant at the 1-percent level. * = statistically significant at the 5-percent level. [†] = statistically significant at the 10-percent level.

Table C.13
Effect of Deployment on Personal Stress, Second-Term-Plus Survey Respondents, with Controls for Overtime and Time Away More or Less than Expected

Variable	Army	Navy	Marine Corps	Air Force
Nonhostile deployment only	0.041**	0.020	0.010	0.038*
	(0.014)	(0.017)	(0.019)	(0.017)
Hostile deployment	0.007	−0.014	−0.025	−0.017
	(0.013)	(0.014)	(0.020)	(0.015)
Away from home but not deployed	−0.016	−0.002	0.004	0.017
	(0.011)	(0.011)	(0.015)	(0.011)
21–60 overtime days	0.077**	0.042**	0.069**	0.046**
	(0.013)	(0.013)	(0.019)	(0.014)
61–120 overtime days	0.131**	0.083**	0.087**	0.077**
	(0.014)	(0.016)	(0.021)	(0.016)
More than 120 overtime days	0.177**	0.120**	0.142**	0.081**
	(0.012)	(0.014)	(0.018)	(0.014)
Much less time away than expected	−0.048**	0.041*	−0.005	0.016
	(0.017)	(0.018)	(0.022)	(0.017)
Less time away than expected	−0.044**	0.024	−0.004	−0.001
	(0.015)	(0.016)	(0.019)	(0.014)
More time away than expected	0.043**	0.081**	0.064**	0.113**
	(0.013)	(0.015)	(0.022)	(0.018)

Table C.13—Continued

Variable	Army	Navy	Marine Corps	Air Force
Much more time away than expected	0.129**	0.134**	0.119**	0.139**
	(0.014)	(0.019)	(0.028)	(0.025)
Very poorly prepared for job	0.048	0.108[†]	0.176*	0.135*
	(0.041)	(0.055)	(0.072)	(0.057)
Poorly prepared for job	0.045[†]	0.045	0.021	0.030
	(0.025)	(0.033)	(0.046)	(0.029)
Well prepared for job	−0.080**	−0.035*	−0.041[†]	−0.031[†]
	(0.015)	(0.018)	(0.024)	(0.017)
Very well prepared for job	−0.112**	−0.069**	−0.088**	−0.062**
	(0.016)	(0.018)	(0.025)	(0.018)
AFQT category I	0.057[†]	0.022	0.076	−0.025
	(0.030)	(0.030)	(0.048)	(0.040)
AFQT category II	0.061**	0.008	0.084*	−0.061[†]
	(0.019)	(0.021)	(0.035)	(0.032)
AFQT category IIIA	0.042*	−0.011	0.074*	−0.076*
	(0.019)	(0.021)	(0.035)	(0.032)
AFQT category IIIB	0.035*	−0.030	0.055	−0.068*
	(0.018)	(0.020)	(0.035)	(0.032)
Some college	−0.011	−0.025	−0.023	0.040[†]
	(0.014)	(0.026)	(0.031)	(0.022)
College graduate	−0.026	−0.014	0.013	0.029
	(0.017)	(0.023)	(0.034)	(0.029)
Black	−0.045**	−0.041**	−0.026	−0.060**
	(0.011)	(0.013)	(0.017)	(0.013)
Hispanic	−0.001	−0.005	−0.029	0.019
	(0.015)	(0.016)	(0.019)	(0.022)
Other race	−0.007	−0.031[†]	0.015	0.003
	(0.018)	(0.016)	(0.031)	(0.021)
Married	0.041**	0.036**	0.041*	0.009
	(0.011)	(0.012)	(0.017)	(0.013)
Number of observations	12,440	9,681	5,608	8,774

NOTE: Models also include controls for location (rural/urban), dual-service career spouse, gender, pay grade, years of service, survey wave indicators, and one-digit DoD occupation. ** = statistically significant at the 1-percent level. * = statistically significant at the 5-percent level. [†] = statistically significant at the 10-percent level.

Table C.14
Effect of Deployment on Intention to Reenlist, First-Term Survey Respondents, with Controls for Overtime and Time Away More or Less than Expected

Variable	Army	Navy	Marine Corps	Air Force
Nonhostile deployment only	0.050*	0.042[†]	0.008	0.024
	(0.021)	(0.025)	(0.026)	(0.029)
Hostile deployment	−0.051**	−0.000	−0.067**	−0.034[†]
	(0.017)	(0.017)	(0.019)	(0.019)
Away from home but not deployed	0.011	0.021	0.001	0.012
	(0.014)	(0.015)	(0.015)	(0.015)
21–60 overtime days	−0.052**	−0.078**	−0.067**	−0.069**
	(0.017)	(0.018)	(0.019)	(0.017)
61–120 overtime days	−0.076**	−0.107**	−0.103**	−0.050*
	(0.019)	(0.022)	(0.021)	(0.022)
More than 120 overtime days	−0.113**	−0.084**	−0.100**	−0.123**
	(0.016)	(0.018)	(0.018)	(0.019)
Much less time away than expected	−0.017	0.003	0.012	−0.029
	(0.022)	(0.024)	(0.020)	(0.019)
Less time away than expected	0.011	−0.013	0.040[†]	−0.004
	(0.020)	(0.023)	(0.020)	(0.019)
More time away than expected	−0.004	−0.069**	−0.008	−0.044[†]
	(0.017)	(0.020)	(0.020)	(0.025)
Much more time away than expected	−0.077**	−0.146**	−0.037	−0.135**
	(0.017)	(0.022)	(0.023)	(0.031)
Very poorly prepared for job	−0.064*	−0.154**	−0.112**	−0.171**
	(0.027)	(0.040)	(0.039)	(0.051)
Poorly prepared for job	−0.002	−0.048	−0.039	−0.034
	(0.022)	(0.033)	(0.030)	(0.035)
Well prepared for job	0.126**	0.112**	0.081**	0.109**
	(0.016)	(0.019)	(0.020)	(0.020)
Very well prepared for job	0.158**	0.208**	0.150**	0.172**
	(0.018)	(0.021)	(0.021)	(0.022)
AFQT category I	−0.099**	−0.103*	−0.045	−0.047
	(0.037)	(0.048)	(0.049)	(0.048)
AFQT category II	−0.120**	−0.132**	−0.014	0.017
	(0.032)	(0.040)	(0.044)	(0.039)
AFQT category IIIA	−0.109**	−0.087*	0.020	0.007
	(0.032)	(0.041)	(0.045)	(0.040)
AFQT category IIIB	−0.081*	−0.063	0.017	0.036
	(0.033)	(0.040)	(0.045)	(0.041)

Table C.14—Continued

Variable	Army	Navy	Marine Corps	Air Force
Some college	−0.012	0.090†	0.076	−0.005
	(0.033)	(0.053)	(0.060)	(0.021)
College graduate	−0.052*	0.030	−0.068	0.028
	(0.021)	(0.040)	(0.060)	(0.055)
Black	0.046**	0.062**	0.050*	0.047**
	(0.016)	(0.020)	(0.022)	(0.018)
Hispanic	0.016	0.059**	0.053**	−0.018
	(0.017)	(0.021)	(0.019)	(0.026)
Other race	−0.019	0.045*	0.041	0.046†
	(0.021)	(0.020)	(0.028)	(0.024)
Married	0.075**	0.098**	0.073**	0.084**
	(0.014)	(0.018)	(0.017)	(0.016)
Number of observations	6,617	5,259	4,853	5,481

NOTE: Models also include controls for location (rural/urban), dual-service career spouse, gender, pay grade, years of service, survey wave indicators, and one-digit DoD occupation. ** = statistically significant at the 1-percent level. * = statistically significant at the 5-percent level. † = statistically significant at the 10-percent level.

Table C.15
Effect of Deployment on Intention to Reenlist, Second-Term-Plus Survey Respondents, with Controls for Overtime and Time Away More or Less than Expected

Variable	Army	Navy	Marine Corps	Air Force
Nonhostile deployment only	0.005	0.029*	0.024	0.010
	(0.012)	(0.014)	(0.016)	(0.014)
Hostile deployment	−0.007	0.053**	0.026	0.003
	(0.012)	(0.012)	(0.017)	(0.013)
Away from home but not deployed	0.007	0.008	0.015	0.012
	(0.010)	(0.009)	(0.013)	(0.010)
21–60 overtime days	−0.039**	−0.038**	−0.039*	−0.045**
	(0.012)	(0.011)	(0.016)	(0.012)
61–120 overtime days	−0.059**	−0.036**	−0.066**	−0.054**
	(0.013)	(0.014)	(0.018)	(0.014)
More than 120 overtime days	−0.074**	−0.056**	−0.057**	−0.070**
	(0.011)	(0.012)	(0.015)	(0.012)
Much less time away than expected	−0.017	−0.002	−0.030	0.026†
	(0.016)	(0.016)	(0.020)	(0.014)
Less time away than expected	−0.001	0.003	−0.003	0.015
	(0.013)	(0.014)	(0.016)	(0.012)
More time away than expected	−0.034**	−0.039**	−0.059**	−0.052**
	(0.012)	(0.013)	(0.019)	(0.016)

Table C.15—Continued

Variable	Army	Navy	Marine Corps	Air Force
Much more time away than expected	−0.106**	−0.098**	−0.119**	−0.116**
	(0.013)	(0.016)	(0.025)	(0.022)
Very poorly prepared for job	−0.182**	−0.140**	−0.274**	−0.248**
	(0.037)	(0.051)	(0.071)	(0.051)
Poorly prepared for job	−0.035	−0.072*	−0.014	−0.052[†]
	(0.025)	(0.031)	(0.041)	(0.027)
Well prepared for job	0.111**	0.116**	0.076**	0.092**
	(0.014)	(0.017)	(0.022)	(0.015)
Very well prepared for job	0.160**	0.137**	0.106**	0.119**
	(0.015)	(0.017)	(0.022)	(0.016)
AFQT category I	−0.025	−0.130**	0.010	−0.041
	(0.028)	(0.027)	(0.044)	(0.033)
AFQT category II	−0.014	−0.065**	0.004	−0.011
	(0.018)	(0.017)	(0.032)	(0.026)
AFQT category IIIA	0.012	−0.005	0.011	−0.016
	(0.017)	(0.017)	(0.032)	(0.026)
AFQT category IIIB	−0.008	0.003	0.017	−0.033
	(0.017)	(0.016)	(0.032)	(0.026)
Some college	0.013	0.008	0.001	−0.073**
	(0.013)	(0.023)	(0.026)	(0.020)
College graduate	−0.028[†]	−0.012	−0.132**	−0.111**
	(0.016)	(0.020)	(0.033)	(0.026)
Black	−0.004	−0.001	0.002	−0.011
	(0.010)	(0.012)	(0.015)	(0.011)
Hispanic	0.001	0.014	0.016	−0.010
	(0.014)	(0.014)	(0.017)	(0.019)
Other race	0.008	0.028*	−0.026	0.021
	(0.017)	(0.014)	(0.026)	(0.017)
Married	0.034**	0.044**	0.037*	0.034**
	(0.011)	(0.011)	(0.016)	(0.012)
Number of observations	13,699	10,323	6,007	9,833

NOTE: Models also include controls for location (rural/urban), dual-service career spouse, gender, pay grade, years of service, survey wave indicators, and one-digit DoD occupation. ** = statistically significant at the 1-percent level. * = statistically significant at the 5-percent level. [†] = statistically significant at the 10-percent level.

Table C.16
Effect of Deployment on Reenlistment, First-Term Survey Respondents, with Controls for Overtime and Time Away More or Less than Expected

Variable	Army	Navy	Marine Corps	Air Force
Nonhostile deployment only	−0.002	0.072†	0.058	0.134**
	(0.038)	(0.041)	(0.037)	(0.043)
Hostile deployment	−0.002	−0.010	−0.066*	0.023
	(0.028)	(0.028)	(0.028)	(0.029)
Away from home but not deployed	0.015	0.026	−0.035	−0.016
	(0.024)	(0.026)	(0.024)	(0.025)
21–60 overtime days	−0.063*	0.012	−0.065*	−0.009
	(0.028)	(0.030)	(0.029)	(0.027)
61–120 Oovertime days	−0.100**	0.016	−0.068*	−0.025
	(0.031)	(0.035)	(0.031)	(0.035)
More than 120 overtime days	−0.077**	0.027	−0.027	−0.041
	(0.027)	(0.030)	(0.028)	(0.031)
Much less time away than expected	−0.034	−0.033	−0.024	0.005
	(0.035)	(0.041)	(0.031)	(0.031)
Less time away than expected	0.028	−0.001	−0.061*	−0.029
	(0.032)	(0.039)	(0.030)	(0.031)
More time away than expected	−0.031	−0.013	−0.066*	0.003
	(0.029)	(0.031)	(0.029)	(0.038)
Much more time away than expected	−0.079**	−0.115**	−0.063†	−0.069
	(0.031)	(0.034)	(0.033)	(0.050)
Very poorly prepared for job	−0.033	−0.176*	−0.169**	−0.105
	(0.058)	(0.084)	(0.057)	(0.089)
Poorly prepared for job	−0.042	−0.097	−0.036	0.051
	(0.041)	(0.061)	(0.052)	(0.056)
Well prepared for job	−0.001	0.103**	0.025	0.072*
	(0.028)	(0.033)	(0.031)	(0.033)
Very well prepared for job	0.030	0.172**	0.038	0.088*
	(0.030)	(0.036)	(0.033)	(0.036)
AFQT category I	−0.062	0.094	−0.023	−0.231**
	(0.063)	(0.093)	(0.080)	(0.080)
AFQT category II	−0.054	−0.025	−0.018	−0.171**
	(0.050)	(0.061)	(0.070)	(0.060)
AFQT category IIIA	−0.048	0.036	0.019	−0.175**
	(0.051)	(0.061)	(0.072)	(0.060)
AFQT category IIIB	−0.049	0.090	−0.005	−0.081
	(0.051)	(0.061)	(0.072)	(0.061)

Table C.16—Continued

Variable	Army	Navy	Marine Corps	Air Force
Some college	−0.000	0.299**	−0.042	0.000
	(0.057)	(0.070)	(0.098)	(0.030)
College graduate	−0.124**	−0.186*	−0.185*	0.124[†]
	(0.038)	(0.092)	(0.086)	(0.075)
Black	0.112**	0.137**	0.158**	0.114**
	(0.025)	(0.031)	(0.033)	(0.028)
Hispanic	0.038	0.062*	0.061*	0.115**
	(0.028)	(0.031)	(0.027)	(0.037)
Other race	0.055	0.050	0.086*	0.079*
	(0.038)	(0.033)	(0.042)	(0.039)
Married	0.124**	0.107**	0.071**	0.090**
	(0.024)	(0.029)	(0.025)	(0.028)
Number of observations	2,603	1,907	2,177	2,048

NOTE: Models also include controls for location (rural/urban), dual-service career spouse, gender, pay grade, years of service, survey wave indicators, and one-digit DoD occupation. ** = statistically significant at the 1-percent level. * = statistically significant at the 5-percent level. [†] = statistically significant at the 10-percent level.

Table C.17
Effect of Deployment on Reenlistment, Second-Term-Plus Survey Respondents, with Controls for Overtime and Time Away More or Less than Expected

Variable	Army	Navy	Marine Corps	Air Force
Nonhostile deployment only	0.031	0.105**	0.107**	0.052*
	(0.021)	(0.026)	(0.025)	(0.022)
Hostile deployment	0.006	0.090**	0.082**	0.056**
	(0.021)	(0.022)	(0.028)	(0.020)
Away from home but not deployed	0.023	0.041*	0.022	0.038*
	(0.017)	(0.017)	(0.021)	(0.015)
21–60 overtime days	−0.027	−0.005	−0.024	−0.011
	(0.021)	(0.021)	(0.027)	(0.019)
61–120 overtime days	0.006	0.066**	−0.028	−0.016
	(0.022)	(0.024)	(0.029)	(0.022)
More than 120 overtime days	−0.024	0.032	0.002	0.016
	(0.020)	(0.021)	(0.025)	(0.020)
Much less time away than expected	−0.059*	−0.100**	−0.054[†]	0.024
	(0.027)	(0.027)	(0.031)	(0.023)
Less time away than expected	0.027	−0.042[†]	−0.012	0.017
	(0.022)	(0.025)	(0.026)	(0.020)
More time away than expected	0.030	−0.008	−0.024	−0.013
	(0.021)	(0.024)	(0.031)	(0.024)

Table C.17—Continued

Variable	Army	Navy	Marine Corps	Air Force
Much more time away than expected	−0.007	−0.009	0.020	−0.000
	(0.023)	(0.029)	(0.038)	(0.034)
Very poorly prepared for job	−0.209**	−0.177*	−0.434**	−0.191*
	(0.065)	(0.085)	(0.094)	(0.088)
Poorly prepared for job	−0.066[†]	−0.050	−0.027	−0.050
	(0.040)	(0.051)	(0.065)	(0.042)
Well prepared for job	0.046[†]	0.041	0.039	0.026
	(0.024)	(0.027)	(0.035)	(0.024)
Very well prepared for job	0.065**	0.019	0.041	0.015
	(0.024)	(0.028)	(0.036)	(0.025)
AFQT category I	0.082[†]	−0.004	0.135[†]	0.124*
	(0.050)	(0.048)	(0.073)	(0.053)
AFQT category II	0.035	0.014	0.112*	0.179**
	(0.029)	(0.030)	(0.054)	(0.043)
AFQT category IIIA	0.110**	0.052[†]	0.105*	0.165**
	(0.028)	(0.031)	(0.053)	(0.043)
AFQT category IIIB	0.092**	0.051[†]	0.109*	0.153**
	(0.026)	(0.029)	(0.052)	(0.043)
Some college	−0.005	−0.012	−0.005	−0.096*
	(0.026)	(0.040)	(0.042)	(0.043)
College graduate	−0.056[†]	−0.002	−0.175**	−0.111*
	(0.032)	(0.035)	(0.046)	(0.050)
Black	0.055**	0.020	0.054*	0.026
	(0.018)	(0.021)	(0.024)	(0.017)
Hispanic	0.026	0.059*	0.055*	−0.014
	(0.022)	(0.025)	(0.027)	(0.031)
Other race	0.001	0.034	0.081*	−0.006
	(0.028)	(0.025)	(0.039)	(0.029)
Married	0.077**	0.014	0.036	−0.010
	(0.020)	(0.020)	(0.027)	(0.020)
Number of observations	4,184	3,870	2,619	3,754

NOTE: Models also include controls for location (rural/urban), dual-service career spouse, gender, pay grade, years of service, survey wave indicators, and one-digit DoD occupation. ** = statistically significant at the 1-percent level. * = statistically significant at the 5-percent level. [†] = statistically significant at the 10-percent level.

Table C.18
Effect of Deployment on Reenlistment in Administrative Data, One-Year Window, First Term and No MOS Controls

Variable	Army	Navy	Marine Corps	Air Force
Nonhostile deployment only	0.100**	0.107**	0.079**	0.184**
	(0.006)	(0.004)	(0.005)	(0.007)
Hostile deployment	−0.041**	−0.021**	−0.019**	−0.022**
	(0.003)	(0.003)	(0.003)	(0.003)
SRB multiplier	−0.009**	0.052**	0.035**	−0.007**
	(0.003)	(0.001)	(0.001)	(0.001)
Years of service	0.024**	−0.088**	−0.175**	−0.017**
	(0.001)	(0.003)	(0.004)	(0.002)
HS dropout or education missing	0.013	0.037**	−0.040*	0.024
	(0.009)	(0.008)	(0.019)	(0.066)
GED	0.035**	0.056**	−0.001	0.100
	(0.006)	(0.011)	(0.013)	(0.069)
At least some college	−0.083**	−0.018†	0.010	−0.107**
	(0.005)	(0.010)	(0.012)	(0.005)
Male	−0.042**	0.020**	−0.028**	−0.019**
	(0.003)	(0.004)	(0.005)	(0.004)
AFQT below category IIIB or missing	−0.043**	0.085**	−0.034*	0.059
	(0.010)	(0.027)	(0.014)	(0.038)
AFQT category IIIB	−0.090**	0.038**	−0.029**	0.024**
	(0.003)	(0.003)	(0.003)	(0.004)
AFQT category II	−0.020**	−0.037**	0.010**	−0.014**
	(0.003)	(0.004)	(0.003)	(0.004)
AFQT category I	−0.055**	−0.079**	−0.011†	−0.033**
	(0.005)	(0.009)	(0.007)	(0.007)
White	−0.019**	−0.062**	−0.027**	−0.037**
	(0.003)	(0.003)	(0.003)	(0.005)
Black	0.104**	0.093**	0.106**	0.069**
	(0.004)	(0.004)	(0.005)	(0.006)
Promoted rapidly	0.303**	0.320**	0.185**	0.187**
	(0.002)	(0.003)	(0.003)	(0.005)
Number of observations	157,319	110,880	115,918	100,848

NOTE: Models also control for year of decision. ** = statistically significant at the 1-percent level. * = statistically significant at the 5-percent level. † = statistically significant at the 10-percent level.

Table C.19
Effect of Deployment on Reenlistment in Administrative Data, One-Year Window, Second Term and No MOS Controls

Variable	Army	Navy	Marine Corps	Air Force
Nonhostile deployment only	0.102**	0.171**	0.115**	0.080**
	(0.005)	(0.005)	(0.008)	(0.007)
Hostile deployment	0.006*	0.039**	0.036**	0.018**
	(0.003)	(0.004)	(0.007)	(0.005)
SRB multiplier	0.007**	−0.016**	0.009*	−0.009**
	(0.003)	(0.001)	(0.004)	(0.001)
Years of service	−0.042**	0.049**	−0.020**	−0.002
	(0.001)	(0.001)	(0.002)	(0.001)
HS dropout or education missing	0.012	0.046**	−0.199†	−0.206
	(0.011)	(0.015)	(0.120)	(0.171)
GED	0.016*	0.013	0.028	0.000
	(0.008)	(0.012)	(0.021)	—
At least some college	−0.012†	0.002	−0.027†	−0.099**
	(0.007)	(0.009)	(0.015)	(0.005)
Male	0.030**	0.070**	0.091**	0.041**
	(0.004)	(0.005)	(0.011)	(0.005)
AFQT below category IIIB or missing	0.045**	0.019	0.016	0.047
	(0.008)	(0.018)	(0.029)	(0.036)
AFQT category IIIB	0.008*	0.037**	0.012†	0.021**
	(0.004)	(0.005)	(0.007)	(0.005)
AFQT category II	−0.020**	−0.067**	−0.028**	0.000
	(0.004)	(0.005)	(0.007)	(0.005)
AFQT category I	−0.061**	−0.164**	−0.053**	−0.013
	(0.009)	(0.008)	(0.017)	(0.011)
White	−0.004	−0.040**	−0.034**	−0.029**
	(0.004)	(0.004)	(0.007)	(0.006)
Black	0.061**	0.034**	0.048**	0.020**
	(0.004)	(0.005)	(0.009)	(0.007)
Promoted rapidly	0.191**	0.273**	0.252**	0.273**
	(0.003)	(0.004)	(0.006)	(0.009)
Number of observations	98,243	75,476	27,833	46,032

NOTE: Models also control for year of decision. ** = statistically significant at the 1-percent level. * = statistically significant at the 5-percent level. † = statistically significant at the 10-percent level.

Table C.20
Effect of Deployment on Reenlistment in Administrative Data, One-Year Window, First Term and MOS Fixed Effects

Variable	Army	Navy	Marine Corps	Air Force
Nonhostile deployment only	0.093**	0.104**	0.086**	0.177**
	(0.006)	(0.004)	(0.005)	(0.007)
Hostile deployment	−0.009**	−0.009**	0.015**	−0.004
	(0.003)	(0.003)	(0.003)	(0.004)
SRB multiplier	0.013**	0.065**	0.078**	0.011**
	(0.003)	(0.002)	(0.001)	(0.002)
Years of service	−0.012**	−0.114**	−0.257**	−0.017**
	(0.002)	(0.003)	(0.006)	(0.002)
HS dropout or education missing	0.020*	0.040**	−0.023	0.003
	(0.009)	(0.008)	(0.018)	(0.065)
GED	0.067**	0.084**	0.015	0.115†
	(0.006)	(0.011)	(0.013)	(0.067)
At least some college	−0.109**	−0.044**	0.001	−0.122**
	(0.005)	(0.010)	(0.011)	(0.005)
Male	0.027**	0.033**	0.004	0.024**
	(0.003)	(0.004)	(0.005)	(0.004)
AFQT below category IIIB or missing	−0.016	0.096**	0.005	0.041
	(0.010)	(0.026)	(0.014)	(0.037)
AFQT category IIIB	−0.087**	0.052**	−0.001	0.022**
	(0.003)	(0.003)	(0.003)	(0.004)
AFQT category II	−0.029**	−0.040**	−0.018**	−0.030**
	(0.003)	(0.004)	(0.003)	(0.004)
AFQT category I	−0.074**	−0.091**	−0.043**	−0.074**
	(0.005)	(0.009)	(0.006)	(0.008)
White	−0.005†	−0.054**	−0.017**	−0.027**
	(0.003)	(0.003)	(0.003)	(0.005)
Black	0.084**	0.102**	0.088**	0.063**
	(0.004)	(0.004)	(0.005)	(0.006)
Promoted rapidly	0.307**	0.321**	0.185**	0.186**
	(0.002)	(0.003)	(0.003)	(0.004)
Sample size	157,319	110,880	115,918	100,848

NOTE: Models also control for year of decision. ** = statistically significant at the 1-percent level. * = statistically significant at the 5-percent level. † = statistically significant at the 10-percent level.

Table C.21
Effect of Deployment on Reenlistment in Administrative Data, One-Year Window, Second Term and MOS Fixed Effects

Variable	Army	Navy	Marine Corps	Air Force
Nonhostile deployment only	0.102**	0.174**	0.126**	0.083**
	(0.005)	(0.005)	(0.008)	(0.007)
Hostile deployment	0.016**	0.042**	0.066**	0.027**
	(0.003)	(0.004)	(0.007)	(0.005)
SRB multiplier	0.025**	0.016**	0.004	0.019**
	(0.003)	(0.003)	(0.005)	(0.003)
Years of service	−0.045**	0.044**	−0.026**	0.001
	(0.001)	(0.001)	(0.002)	(0.001)
HS dropout or education missing	0.016	0.032*	−0.217†	−0.207
	(0.011)	(0.015)	(0.114)	(0.174)
GED	0.033**	0.012	0.026	0.000
	(0.008)	(0.012)	(0.021)	—
At least some college	−0.024**	−0.008	−0.027†	−0.103**
	(0.007)	(0.009)	(0.015)	(0.005)
Male	0.047**	0.065**	0.092**	0.060**
	(0.004)	(0.005)	(0.011)	(0.005)
AFQT below category IIIB or missing	0.060**	0.033†	0.028	0.035
	(0.009)	(0.018)	(0.029)	(0.036)
AFQT category IIIB	0.015**	0.035**	0.014†	0.016**
	(0.004)	(0.005)	(0.007)	(0.005)
AFQT category II	−0.023**	−0.044**	−0.029**	−0.004
	(0.004)	(0.005)	(0.007)	(0.005)
AFQT category I	−0.065**	−0.102**	−0.052**	−0.025*
	(0.009)	(0.008)	(0.017)	(0.011)
White	0.001	−0.041**	−0.026**	−0.024**
	(0.004)	(0.004)	(0.007)	(0.006)
Black	0.058**	0.041**	0.050**	0.018*
	(0.004)	(0.005)	(0.009)	(0.007)
Promoted rapidly	0.195**	0.260**	0.233**	0.267**
	(0.003)	(0.004)	(0.006)	(0.009)
Number of observations	98,243	75,476	27,833	46,032

NOTE: Models also control for year of decision. ** = statistically significant at the 1-percent level. * = statistically significant at the 5-percent level. † = statistically significant at the 10-percent level.

Table C.22
Effect of Deployment on Reenlistment in Administrative Data, One-Year Window, First Term and MOS-by-Quarter Fixed Effects

Variable	Army	Navy	Marine Corps	Air Force
Nonhostile deployment only	0.082**	0.097**	0.065**	0.164**
	(0.006)	(0.004)	(0.004)	(0.007)
Hostile deployment	−0.005†	−0.004	0.017**	0.000
	(0.003)	(0.003)	(0.002)	(0.004)
Years of service	−0.004**	−0.102**	−0.193**	−0.008**
	(0.001)	(0.003)	(0.006)	(0.002)
HS dropout or education missing	0.020*	0.020*	−0.025	−0.013
	(0.009)	(0.008)	(0.019)	(0.062)
GED	0.062**	0.067**	0.012	0.101
	(0.006)	(0.011)	(0.011)	(0.069)
At least some college	−0.110**	−0.047**	−0.015	−0.116**
	(0.005)	(0.010)	(0.010)	(0.005)
Male	0.030**	0.033**	0.006	0.026**
	(0.003)	(0.004)	(0.005)	(0.004)
AFQT below category IIIB or missing	−0.016	0.091**	−0.015	0.054
	(0.010)	(0.026)	(0.012)	(0.037)
AFQT category IIIB	−0.084**	0.047**	−0.006*	0.024**
	(0.003)	(0.003)	(0.003)	(0.004)
AFQT category II	−0.028**	−0.039**	−0.014**	−0.027**
	(0.003)	(0.004)	(0.003)	(0.004)
AFQT category I	−0.068**	−0.093**	−0.042**	−0.071**
	(0.005)	(0.009)	(0.006)	(0.008)
White	−0.007*	−0.051**	−0.008**	−0.023**
	(0.003)	(0.003)	(0.003)	(0.005)
Black	0.081**	0.097**	0.078**	0.061**
	(0.004)	(0.004)	(0.004)	(0.005)
Promoted rapidly	0.298**	0.305**	0.151**	0.185**
	(0.002)	(0.003)	(0.002)	(0.004)
Number of observations	157,759	111,102	123,565	101,140

NOTE: Models also control for year of decision. ** = statistically significant at the 1-percent level. * = statistically significant at the 5-percent level. † = statistically significant at the 10-percent level.

Table C.23
Effect of Deployment on Reenlistment in Administrative Data, One-Year Window, Second Term and MOS-by-Quarter Fixed Effects

Variable	Army	Navy	Marine Corps	Air Force
Nonhostile deployment only	0.096**	0.169**	0.124**	0.080**
	(0.005)	(0.005)	(0.008)	(0.007)
Hostile deployment	0.014**	0.038**	0.072**	0.028**
	(0.003)	(0.004)	(0.007)	(0.005)
Years of service	−0.036**	0.007**	−0.023**	−0.009**
	(0.001)	(0.001)	(0.002)	(0.001)
HS dropout or education missing	0.015	−0.002	−0.219†	−0.197
	(0.011)	(0.014)	(0.113)	(0.158)
GED	0.030**	−0.007	0.020	0.175**
	(0.008)	(0.011)	(0.021)	(0.050)
At least some college	−0.020**	−0.011	−0.022	−0.099**
	(0.007)	(0.009)	(0.015)	(0.005)
Male	0.046**	0.072**	0.093**	0.060**
	(0.004)	(0.005)	(0.012)	(0.005)
AFQT below category IIIB or missing	0.065**	−0.068**	0.031	−0.064*
	(0.008)	(0.015)	(0.028)	(0.028)
AFQT category IIIB	0.016**	0.028**	0.012†	0.010†
	(0.004)	(0.005)	(0.007)	(0.005)
AFQT category II	−0.022**	−0.043**	−0.027**	−0.003
	(0.004)	(0.005)	(0.007)	(0.005)
AFQT category I	−0.066**	−0.101**	−0.052**	−0.026*
	(0.009)	(0.008)	(0.017)	(0.011)
White	0.002	−0.043**	−0.026**	−0.023**
	(0.004)	(0.004)	(0.007)	(0.006)
Black	0.058**	0.046**	0.047**	0.018*
	(0.004)	(0.005)	(0.009)	(0.007)
Promoted rapidly	0.186**	0.246**	0.220**	0.266**
	(0.003)	(0.004)	(0.006)	(0.008)
Number of observations	99,082	79,270	28,478	47,647

NOTE: Models also control for year of decision. ** = statistically significant at the 1-percent level. * = statistically significant at the 5-percent level. † = statistically significant at the 10-percent level.

Table C.24
Effect of Deployment on Reenlistment in Administrative Data, Three-Year Window, First Term and No MOS Controls

Variable	Army	Navy	Marine Corps	Air Force
Nonhostile deployment only	0.141**	0.106**	0.090**	0.179**
	(0.005)	(0.006)	(0.005)	(0.006)
Hostile deployment	−0.032**	−0.022**	−0.029**	−0.016**
	(0.003)	(0.003)	(0.003)	(0.003)
SRB multiplier	−0.007**	0.052**	0.036**	−0.006**
	(0.003)	(0.001)	(0.001)	(0.001)
Years of service	0.024**	−0.089**	−0.174**	−0.016**
	(0.001)	(0.003)	(0.004)	(0.002)
HS dropout or education missing	0.012	0.037**	−0.037*	0.030
	(0.009)	(0.008)	(0.019)	(0.066)
GED	0.036**	0.058**	0.005	0.103
	(0.006)	(0.011)	(0.013)	(0.069)
At least some college	−0.088**	−0.019†	0.004	−0.110**
	(0.005)	(0.010)	(0.012)	(0.005)
Male	−0.041**	0.026**	−0.023**	−0.016**
	(0.003)	(0.004)	(0.005)	(0.004)
AFQT below category IIIB or missing	−0.048**	0.087**	−0.031*	0.055
	(0.010)	(0.027)	(0.014)	(0.038)
AFQT category IIIB	−0.092**	0.041**	−0.028**	0.025**
	(0.003)	(0.003)	(0.003)	(0.004)
AFQT category II	−0.020**	−0.039**	0.009**	−0.016**
	(0.003)	(0.004)	(0.003)	(0.004)
AFQT category I	−0.056**	−0.084**	−0.013†	−0.037**
	(0.005)	(0.009)	(0.007)	(0.007)
White	−0.017**	−0.064**	−0.027**	−0.036**
	(0.003)	(0.003)	(0.003)	(0.005)
Black	0.105**	0.093**	0.104**	0.071**
	(0.004)	(0.004)	(0.005)	(0.006)
Promoted rapidly	0.299**	0.321**	0.185**	0.187**
	(0.002)	(0.003)	(0.003)	(0.005)
Number of observations	157,319	110,880	115,918	100,848

NOTE: Models also control for year of decision. ** = statistically significant at the 1-percent level. * = statistically significant at the 5-percent level. † = statistically significant at the 10-percent level.

Table C.25
Effect of Deployment on Reenlistment in Administrative Data, Three-Year Window, Second Term and No MOS Controls

Variable	Army	Navy	Marine Corps	Air Force
Nonhostile deployment only	0.088**	0.153**	0.131**	0.061**
	(0.005)	(0.005)	(0.008)	(0.006)
Hostile deployment	0.042**	0.078**	0.059**	0.020**
	(0.004)	(0.004)	(0.007)	(0.004)
SRB multiplier	0.007*	−0.016**	0.011**	−0.010**
	(0.003)	(0.001)	(0.004)	(0.001)
Years of service	−0.042**	0.050**	−0.020**	−0.002
	(0.001)	(0.001)	(0.002)	(0.001)
HS dropout or education missing	0.010	0.044**	−0.185	−0.202
	(0.011)	(0.015)	(0.119)	(0.171)
GED	0.014†	0.010	0.027	0.000
	(0.008)	(0.012)	(0.021)	.
At least some college	−0.011	0.003	−0.027†	−0.099**
	(0.007)	(0.009)	(0.015)	(0.005)
Male	0.025**	0.065**	0.087**	0.041**
	(0.004)	(0.005)	(0.011)	(0.005)
AFQT below category IIIB or missing	0.043**	0.019	0.016	0.044
	(0.008)	(0.018)	(0.029)	(0.036)
AFQT category IIIB	0.007*	0.034**	0.011	0.021**
	(0.004)	(0.005)	(0.007)	(0.005)
AFQT category II	−0.020**	−0.067**	−0.027**	−0.000
	(0.004)	(0.005)	(0.007)	(0.005)
AFQT category I	−0.060**	−0.163**	−0.052**	−0.013
	(0.009)	(0.008)	(0.017)	(0.011)
White	−0.005	−0.039**	−0.033**	−0.029**
	(0.004)	(0.004)	(0.007)	(0.006)
Black	0.061**	0.033**	0.049**	0.020**
	(0.004)	(0.005)	(0.009)	(0.007)
Promoted rapidly	0.190**	0.269**	0.248**	0.273**
	(0.003)	(0.004)	(0.006)	(0.009)
Number of observations	98,243	75,476	27,833	46,032

NOTE: Models also control for year of decision. ** = statistically significant at the 1-percent level. * = statistically significant at the 5-percent level. † = statistically significant at the 10-percent level.

**Table C.26
Effect of Deployment on Reenlistment in Administrative Data, Three-Year Window, First Term and MOS Fixed Effects**

Variable	Army	Navy	Marine Corps	Air Force
Nonhostile deployment only	0.139**	0.108**	0.084**	0.173**
	(0.005)	(0.005)	(0.005)	(0.006)
Hostile deployment	0.002	0.011**	0.007*	0.008*
	(0.003)	(0.003)	(0.003)	(0.003)
SRB multiplier	0.014**	0.066**	0.078**	0.011**
	(0.003)	(0.002)	(0.001)	(0.002)
Years of service	−0.013**	−0.114**	−0.257**	−0.017**
	(0.002)	(0.003)	(0.006)	(0.002)
HS dropout or education missing	0.020*	0.040**	−0.023	0.011
	(0.009)	(0.008)	(0.018)	(0.065)
GED	0.068**	0.084**	0.019	0.117†
	(0.006)	(0.011)	(0.013)	(0.067)
At least some college	−0.113**	−0.043**	−0.003	−0.124**
	(0.005)	(0.010)	(0.011)	(0.005)
Male	0.027**	0.034**	0.007	0.025**
	(0.003)	(0.004)	(0.005)	(0.004)
AFQT below category IIIB or missing	−0.020*	0.097**	0.006	0.037
	(0.010)	(0.027)	(0.014)	(0.036)
AFQT category IIIB	−0.088**	0.052**	−0.001	0.022**
	(0.003)	(0.003)	(0.003)	(0.004)
AFQT category II	−0.030**	−0.040**	−0.018**	−0.031**
	(0.003)	(0.004)	(0.003)	(0.004)
AFQT category I	−0.073**	−0.092**	−0.043**	−0.076**
	(0.005)	(0.009)	(0.006)	(0.008)
White	−0.004	−0.055**	−0.018**	−0.026**
	(0.003)	(0.003)	(0.003)	(0.005)
Black	0.086**	0.102**	0.087**	0.064**
	(0.004)	(0.004)	(0.005)	(0.006)
Promoted rapidly	0.304**	0.322**	0.186**	0.186**
	(0.002)	(0.003)	(0.003)	(0.004)
Number of observations	157,319	110,880	115,918	100,848

NOTE: Models also control for year of decision. ** = statistically significant at the 1-percent level. * = statistically significant at the 5-percent level. † = statistically significant at the 10-percent level.

Table C.27
Effect of Deployment on Reenlistment in Administrative Data, Three-Year Window, Second Term and MOS Fixed Effects

Variable	Army	Navy	Marine Corps	Air Force
Nonhostile deployment only	0.089**	0.154**	0.127**	0.065**
	(0.005)	(0.005)	(0.008)	(0.006)
Hostile deployment	0.055**	0.081**	0.079**	0.032**
	(0.004)	(0.004)	(0.007)	(0.004)
SRB multiplier	0.025**	0.016**	0.005	0.019**
	(0.003)	(0.003)	(0.005)	(0.003)
Years of service	−0.045**	0.045**	−0.025**	0.001
	(0.001)	(0.001)	(0.002)	(0.001)
HS dropout or education missing	0.015	0.030*	−0.201†	−0.203
	(0.011)	(0.015)	(0.114)	(0.174)
GED	0.033**	0.010	0.026	0.000
	(0.008)	(0.012)	(0.021)	—
At least some college	−0.024**	−0.008	−0.028†	−0.103**
	(0.007)	(0.009)	(0.015)	(0.005)
Male	0.043**	0.061**	0.089**	0.060**
	(0.004)	(0.005)	(0.011)	(0.005)
AFQT below category IIIB or missing	0.059**	0.034†	0.027	0.032
	(0.009)	(0.018)	(0.029)	(0.036)
AFQT category IIIB	0.015**	0.034**	0.014†	0.016**
	(0.004)	(0.005)	(0.007)	(0.005)
AFQT category II	−0.023**	−0.044**	−0.029**	−0.004
	(0.004)	(0.005)	(0.007)	(0.005)
AFQT category I	−0.065**	−0.101**	−0.052**	−0.025*
	(0.009)	(0.008)	(0.017)	(0.011)
White	0.000	−0.040**	−0.025**	−0.024**
	(0.004)	(0.004)	(0.007)	(0.006)
Black	0.058**	0.041**	0.050**	0.018*
	(0.004)	(0.005)	(0.009)	(0.007)
Promoted rapidly	0.195**	0.257**	0.231**	0.267**
	(0.003)	(0.004)	(0.006)	(0.009)
Number of observations	98,243	75,476	27,833	46,032

NOTE: Models also control for year of decision. ** = statistically significant at the 1-percent level. * = statistically significant at the 5-percent level. † = statistically significant at the 10-percent level.

Table C.28
Effect of Deployment on Reenlistment in Administrative Data, Three-Year Window, First Term and MOS-by-Quarter Fixed Effects

Variable	Army	Navy	Marine Corps	Air Force
Nonhostile deployment only	0.128**	0.099**	0.060**	0.160**
	(0.005)	(0.005)	(0.004)	(0.006)
Hostile deployment	0.004	0.013**	0.000	0.009**
	(0.003)	(0.003)	(0.003)	(0.003)
Years of service	−0.005**	−0.102**	−0.193**	−0.008**
	(0.001)	(0.003)	(0.006)	(0.002)
HS dropout or education missing	0.019*	0.019*	−0.024	−0.006
	(0.009)	(0.008)	(0.019)	(0.062)
GED	0.064**	0.067**	0.015	0.103
	(0.006)	(0.011)	(0.011)	(0.069)
At least some college	−0.114**	−0.047**	−0.017†	−0.118**
	(0.005)	(0.010)	(0.010)	(0.005)
Male	0.029**	0.035**	0.010*	0.026**
	(0.003)	(0.004)	(0.005)	(0.004)
AFQT below category IIIB or missing	−0.019†	0.091**	−0.015	0.051
	(0.010)	(0.026)	(0.012)	(0.037)
AFQT category IIIB	−0.084**	0.047**	−0.006*	0.024**
	(0.003)	(0.003)	(0.003)	(0.004)
AFQT category II	−0.028**	−0.039**	−0.014**	−0.028**
	(0.003)	(0.004)	(0.003)	(0.004)
AFQT category I	−0.068**	−0.094**	−0.042**	−0.072**
	(0.005)	(0.009)	(0.006)	(0.008)
White	−0.005†	−0.052**	−0.008**	−0.023**
	(0.003)	(0.003)	(0.003)	(0.005)
Black	0.082**	0.097**	0.077**	0.062**
	(0.004)	(0.004)	(0.004)	(0.005)
Promoted rapidly	0.295**	0.305**	0.153**	0.185**
	(0.002)	(0.003)	(0.002)	(0.004)
Number of observations	157,759	111,102	123,565	101,140

NOTE: Models also control for year of decision. ** = statistically significant at the 1-percent level. * = statistically significant at the 5-percent level. † = statistically significant at the 10-percent level.

Table C.29
Effect of Deployment on Reenlistment in Administrative Data, Three-Year Window, Second Term and MOS-by-Quarter Fixed Effects

Variable	Army	Navy	Marine Corps	Air Force
Nonhostile deployment only	0.085**	0.149**	0.123**	0.066**
	(0.005)	(0.005)	(0.008)	(0.006)
Hostile deployment	0.052**	0.070**	0.086**	0.035**
	(0.004)	(0.004)	(0.007)	(0.004)
Years of service	−0.036**	0.008**	−0.023**	−0.009**
	(0.001)	(0.001)	(0.002)	(0.001)
HS dropout or education missing	0.014	−0.004	−0.202†	−0.189
	(0.011)	(0.014)	(0.112)	(0.158)
GED	0.030**	−0.009	0.019	0.183**
	(0.008)	(0.011)	(0.021)	(0.047)
At least some college	−0.020**	−0.010	−0.023	−0.099**
	(0.007)	(0.009)	(0.015)	(0.005)
Male	0.042**	0.069**	0.090**	0.060**
	(0.004)	(0.005)	(0.012)	(0.005)
AFQT below category IIIB or missing	0.064**	−0.068**	0.032	−0.066*
	(0.008)	(0.015)	(0.028)	(0.028)
AFQT category IIIB	0.015**	0.027**	0.012	0.010†
	(0.004)	(0.005)	(0.007)	(0.005)
AFQT category II	−0.022**	−0.043**	−0.027**	−0.003
	(0.004)	(0.005)	(0.007)	(0.005)
AFQT category I	−0.065**	−0.100**	−0.052**	−0.025*
	(0.009)	(0.008)	(0.017)	(0.011)
White	0.001	−0.042**	−0.025**	−0.023**
	(0.004)	(0.004)	(0.007)	(0.006)
Black	0.057**	0.046**	0.048**	0.018*
	(0.004)	(0.005)	(0.009)	(0.007)
Promoted rapidly	0.186**	0.244**	0.217**	0.266**
	(0.003)	(0.004)	(0.006)	(0.008)
Number of observations	99,082	79,270	28,478	47,647

NOTE: Models also control for year of decision. ** = statistically significant at the 1-percent level. * = statistically significant at the 5-percent level. † = statistically significant at the 10-percent level.

Table C.30
Effect of Deployment in Prior Year on Reenlistment in Administrative Data, by Year of Decision

Year of Decision	First Term				Second Term			
	Army	Navy	Marine Corps	Air Force	Army	Navy	Marine Corps	Air Force
1996	0.057**	0.016*	0.018*	0.029**	0.093**	0.086**	0.107**	0.065**
	(0.008)	(0.007)	(0.009)	(0.008)	(0.007)	(0.007)	(0.021)	(0.006)
1997	0.127**	0.010†	−0.006	0.014†	0.106**	0.086**	0.151**	0.059**
	(0.007)	(0.006)	(0.008)	(0.008)	(0.007)	(0.008)	(0.024)	(0.008)
1998	0.063**	−0.008	−0.013†	−0.011	0.090**	0.131**	0.166**	0.051**
	(0.008)	(0.007)	(0.008)	(0.008)	(0.009)	(0.009)	(0.027)	(0.008)
1999	0.056**	−0.012	0.005	0.001	0.103**	0.116**	0.124**	0.066**
	(0.007)	(0.008)	(0.007)	(0.008)	(0.009)	(0.009)	(0.027)	(0.009)
2000	0.078**	−0.008	0.001	−0.003	0.104**	0.062**	0.088**	0.070**
	(0.007)	(0.007)	(0.007)	(0.008)	(0.009)	(0.009)	(0.025)	(0.009)
2001	0.047**	−0.055**	−0.031**	−0.015	0.093**	0.032**	0.086**	0.057**
	(0.007)	(0.008)	(0.008)	(0.009)	(0.010)	(0.010)	(0.029)	(0.011)
2002	0.023**	−0.030**	0.001	−0.009	0.055**	0.018†	0.064**	0.024**
	(0.007)	(0.007)	(0.007)	(0.008)	(0.010)	(0.010)	(0.024)	(0.010)
2003	0.043**	−0.005	0.004	−0.024**	0.073**	−0.005	0.013	0.010
	(0.006)	(0.007)	(0.005)	(0.010)	(0.009)	(0.010)	(0.019)	(0.011)
2004	0.022**	0.016	−0.000	−0.016†	0.081**	0.027**	0.056**	0.014
	(0.006)	(0.007)	(0.006)	(0.009)	(0.008)	(0.010)	(0.017)	(0.012)
2005	0.019**	−0.013†	0.005	−0.003	0.005	0.068**	0.067**	0.039**
	(0.006)	(0.007)	(0.006)	(0.008)	(0.007)	(0.010)	(0.016)	(0.012)
2006	−0.082**	0.015*	0.023**	0.015	−0.083**	0.064**	0.106**	0.045**
	(0.006)	(0.007)	(0.005)	(0.008)	(0.007)	(0.009)	(0.015)	(0.011)
2007	−0.051**	−0.008	0.063**	0.023**	−0.000	0.050**	0.102**	0.038**
	(0.008)	(0.008)	(0.007)	(0.009)	(0.008)	(0.011)	(0.016)	(0.012)

NOTE: Models include controls for nonhostile deployment only in window, years of service at the time of the decision, education, gender, AFQT, race, and an indicator for being promoted more rapidly than is typical.
** = statistically significant at the 1-percent level. * = statistically significant at the 5-percent level. † = statistically significant at the 10-percent level.

Table C.31
Effect of Deployments in Prior Year and First Two Years of Three-Year Window, by Year of Decision, Army

Year of Decision	First Term			Second Term		
	Deployed Prior Year Only	Deployed First 2 Years of Window Only	Deployed Prior Year and First 2 Years of Window	Deployed Prior Year Only	Deployed First 2 Years of Window Only	Deployed Prior Year and First 2 Years of Window
1996	0.080**	0.041**	0.058**	0.126**	0.072**	0.091**
	(0.010)	(0.008)	(0.013)	(0.009)	(0.007)	(0.012)
1997	0.147**	0.013	0.124**	0.136**	0.060**	0.113**
	(0.009)	(0.009)	(0.011)	(0.009)	(0.008)	(0.011)
1998	0.065**	0.067**	0.103**	0.115**	0.056**	0.109**
	(0.011)	(0.009)	(0.011)	(0.013)	(0.010)	(0.013)
1999	0.096**	0.045**	0.052**	0.150**	0.039**	0.081**
	(0.010)	(0.008)	(0.010)	(0.013)	(0.009)	(0.013)
2000	0.115**	0.015*	0.057**	0.158**	0.014	0.051**
	(0.009)	(0.008)	(0.011)	(0.011)	(0.010)	(0.014)
2001	0.060**	0.023**	0.058**	0.158**	0.023**	0.062**
	(0.010)	(0.007)	(0.009)	(0.013)	(0.009)	(0.012)
2002	0.052**	0.009	0.027**	0.098**	0.024**	0.041**
	(0.011)	(0.008)	(0.009)	(0.014)	(0.010)	(0.012)
2003	0.083**	0.008	0.013	0.107**	0.048**	0.041**
	(0.007)	(0.009)	(0.008)	(0.010)	(0.014)	(0.013)
2004	0.116**	−0.027*	−0.042**	0.153**	0.056**	0.030**
	(0.007)	(0.013)	(0.007)	(0.010)	(0.016)	(0.011)
2005	0.112**	−0.020*	−0.022**	0.100**	0.056**	−0.010
	(0.010)	(0.009)	(0.008)	(0.009)	(0.009)	(0.008)
2006	0.014	−0.050**	−0.137**	0.028**	0.060**	−0.070**
	(0.010)	(0.010)	(0.008)	(0.008)	(0.008)	(0.008)
2007	0.121**	−0.039**	−0.117**	0.086**	0.042**	−0.020[†]
	(0.012)	(0.012)	(0.010)	(0.010)	(0.012)	(0.010)

NOTE: Models include controls for nonhostile deployment only in window, years of service at the time of the decision, education, gender, AFQT, race, and an indicator for being promoted more rapidly than is typical.
** = statistically significant at the 1-percent level. * = statistically significant at the 5-percent level. [†] = statistically significant at the 10-percent level.

Table C.32
Effect of Deployments in Prior Year and First Two Years of Three-Year Window, by Year of Decision, Navy

Year of Decision	First Term			Second Term		
	Deployed Prior Year Only	Deployed First 2 Years of Window Only	Deployed Prior Year and First 2 Years of Window	Deployed Prior Year Only	Deployed First 2 Years of Window Only	Deployed Prior Year and First 2 Years of Window
1996	0.024*	0.001	0.001	0.110**	0.080**	0.104**
	(0.011)	(0.008)	(0.009)	(0.010)	(0.008)	(0.010)
1997	0.020*	0.024**	0.003	0.122**	0.117**	0.125**
	(0.009)	(0.007)	(0.009)	(0.011)	(0.009)	(0.012)
1998	−0.008	0.026**	−0.000	0.150**	0.108**	0.176**
	(0.011)	(0.009)	(0.010)	(0.013)	(0.010)	(0.012)
1999	−0.005	−0.002	−0.029**	0.141**	0.087**	0.141**
	(0.011)	(0.010)	(0.010)	(0.013)	(0.011)	(0.012)
2000	−0.007	0.007	−0.014	0.100**	0.087**	0.084**
	(0.011)	(0.010)	(0.010)	(0.013)	(0.010)	(0.012)
2001	−0.055**	0.021*	−0.052**	0.102**	0.086**	0.035**
	(0.013)	(0.009)	(0.011)	(0.016)	(0.010)	(0.013)
2002	−0.011	0.013	−0.039**	0.052**	0.085**	0.046**
	(0.010)	(0.008)	(0.009)	(0.015)	(0.010)	(0.013)
2003	0.011	0.038**	0.006	0.029†	0.049**	0.010
	(0.012)	(0.010)	(0.010)	(0.015)	(0.012)	(0.013)
2004	0.024*	0.028**	0.022**	0.060**	0.057**	0.041**
	(0.012)	(0.009)	(0.009)	(0.016)	(0.011)	(0.012)
2005	−0.018	0.019*	−0.008	0.103**	0.066**	0.086**
	(0.012)	(0.009)	(0.010)	(0.015)	(0.011)	(0.012)
2006	0.026*	0.036**	0.030**	0.106**	0.084**	0.092**
	(0.012)	(0.009)	(0.010)	(0.014)	(0.010)	(0.012)
2007	0.007	0.027**	−0.006	0.077**	0.105**	0.096**
	(0.012)	(0.011)	(0.011)	(0.016)	(0.012)	(0.015)

NOTE: Models include controls for nonhostile deployment only in window, years of service at the time of the decision, education, gender, AFQT, race, and an indicator for being promoted more rapidly than is typical.
** = statistically significant at the 1-percent level. * = statistically significant at the 5-percent level. † = statistically significant at the 10-percent level.

Table C.33
Effect of Deployments in Prior Year and First Two Years of Three-Year Window, by Year of Decision, Marine Corps

Year of Decision	First Term			Second Term		
	Deployed Prior Year Only	Deployed First 2 Years of Window Only	Deployed Prior Year and First 2 Years of Window	Deployed Prior Year Only	Deployed First 2 Years of Window Only	Deployed Prior Year and First 2 Years of Window
1996	0.015	0.013†	0.027*	0.138**	0.091**	0.133**
	(0.011)	(0.007)	(0.013)	(0.027)	(0.017)	(0.033)
1997	0.011	0.011	−0.017	0.247**	0.101**	0.120**
	(0.010)	(0.008)	(0.011)	(0.030)	(0.023)	(0.040)
1998	0.001	0.012	−0.022*	0.182**	0.101**	0.193**
	(0.011)	(0.009)	(0.011)	(0.035)	(0.027)	(0.041)
1999	0.017†	−0.016†	−0.017†	0.166**	0.080**	0.123**
	(0.009)	(0.009)	(0.009)	(0.034)	(0.029)	(0.043)
2000	0.008	0.010	−0.000	0.092**	0.063**	0.116**
	(0.010)	(0.008)	(0.009)	(0.031)	(0.025)	(0.041)
2001	−0.013	0.010	−0.053**	0.146**	0.075**	0.059
	(0.010)	(0.007)	(0.010)	(0.037)	(0.024)	(0.046)
2002	0.009	0.001	−0.012	0.127**	0.084**	0.043
	(0.009)	(0.008)	(0.010)	(0.029)	(0.023)	(0.039)
2003	0.006	−0.020†	−0.008	0.050*	0.025	0.024
	(0.006)	(0.011)	(0.007)	(0.023)	(0.029)	(0.029)
2004	0.028**	−0.024**	−0.028**	0.108**	0.054*	0.052*
	(0.009)	(0.010)	(0.007)	(0.024)	(0.025)	(0.023)
2005	0.028**	−0.048**	−0.036**	0.113**	0.056**	0.096**
	(0.010)	(0.008)	(0.008)	(0.025)	(0.020)	(0.020)
2006	0.069**	0.009	0.013†	0.167**	0.096**	0.151**
	(0.010)	(0.008)	(0.008)	(0.026)	(0.019)	(0.020)
2007	0.094**	−0.016	0.034**	0.180**	0.045*	0.106**
	(0.013)	(0.011)	(0.010)	(0.026)	(0.020)	(0.021)

NOTE: Models include controls for nonhostile deployment only in window, years of service at the time of the decision, education, gender, AFQT, race, and an indicator for being promoted more rapidly than is typical.
** = statistically significant at the 1-percent level. * = statistically significant at the 5-percent level. † = statistically significant at the 10-percent level.

**Table C.34
Effect of Deployments in Prior Year and First Two Years of Three-Year Window, by Year of Decision, Air Force**

Year of Decision	First Term			Second Term		
	Deployed Prior Year Only	Deployed First 2 Years of Window Only	Deployed Prior Year and First 2 Years of Window	Deployed Prior Year Only	Deployed First 2 Years of Window Only	Deployed Prior Year and First 2 Years of Window
1996	0.058**	0.021†	0.029**	0.071**	0.048**	0.087**
	(0.010)	(0.011)	(0.012)	(0.009)	(0.007)	(0.008)
1997	0.047**	−0.002	−0.004	0.097**	0.040**	0.053**
	(0.010)	(0.012)	(0.012)	(0.010)	(0.009)	(0.011)
1998	0.024**	−0.019	−0.035**	0.081**	0.033**	0.047**
	(0.010)	(0.012)	(0.011)	(0.011)	(0.009)	(0.011)
1999	0.037**	−0.008	−0.019†	0.090**	0.039**	0.071**
	(0.011)	(0.012)	(0.011)	(0.013)	(0.010)	(0.012)
2000	0.039**	−0.021†	−0.039**	0.094**	0.025**	0.072**
	(0.011)	(0.011)	(0.011)	(0.012)	(0.010)	(0.012)
2001	0.014	−0.009	−0.035**	0.079**	0.027**	0.056**
	(0.012)	(0.010)	(0.013)	(0.016)	(0.011)	(0.014)
2002	0.004	0.006	−0.014	0.048**	0.035**	0.029*
	(0.011)	(0.010)	(0.011)	(0.014)	(0.011)	(0.013)
2003	−0.004	0.017	−0.030*	0.040**	0.034**	0.010
	(0.013)	(0.012)	(0.013)	(0.015)	(0.013)	(0.015)
2004	0.021	0.036**	−0.017	0.045**	0.015	0.012
	(0.013)	(0.011)	(0.011)	(0.018)	(0.015)	(0.016)
2005	0.024*	0.009	−0.009	0.068**	0.037**	0.045**
	(0.012)	(0.010)	(0.011)	(0.018)	(0.014)	(0.016)
2006	0.033**	0.010	0.020*	0.076**	0.008	0.033*
	(0.011)	(0.009)	(0.010)	(0.016)	(0.013)	(0.015)
2007	0.058**	0.003	0.000	0.049**	0.039**	0.049**
	(0.013)	(0.011)	(0.011)	(0.018)	(0.015)	(0.016)

NOTE: Models include controls for nonhostile deployment only in window, years of service at the time of the decision, education, gender, AFQT, race, and an indicator for being promoted more rapidly than is typical.
** = statistically significant at the 1-percent level. * = statistically significant at the 5-percent level. † = statistically significant at the 10-percent level.

Table C.35
Effect of Months of Deployment in Prior Three Years on Reenlistment, by Year of Decision, Army

Year of Decision	First Term				Second Term			
	1–6 Months	7–11 Months	12–17 Months	18+ Months	1–6 Months	7–11 Months	12–17 Months	18+ Months
1996	0.058**	0.047**	0.041	0.087	0.094**	0.082**	0.040	0.109+
	(0.006)	(0.016)	(0.049)	(0.076)	(0.006)	(0.015)	(0.031)	(0.059)
1997	0.061**	0.176**	0.070**	0.107	0.086**	0.132**	0.110**	0.037
	(0.007)	(0.011)	(0.029)	(0.119)	(0.007)	(0.011)	(0.026)	(0.064)
1998	0.058**	0.117**	0.041	0.007	0.075**	0.105**	0.067**	−0.033
	(0.008)	(0.011)	(0.025)	(0.097)	(0.009)	(0.012)	(0.029)	(0.069)
1999	0.068**	0.053**	−0.002	−0.089	0.066**	0.081**	0.067**	0.054
	(0.007)	(0.009)	(0.026)	(0.068)	(0.009)	(0.012)	(0.025)	(0.062)
2000	0.077**	0.016+	0.009	−0.148*	0.077**	0.043**	−0.005	−0.008
	(0.007)	(0.009)	(0.032)	(0.075)	(0.009)	(0.013)	(0.033)	(0.071)
2001	0.052**	0.020**	−0.009	−0.070	0.075**	0.024*	−0.019	0.089
	(0.006)	(0.008)	(0.025)	(0.093)	(0.008)	(0.012)	(0.030)	(0.112)
2002	0.037**	0.013	−0.046*	−0.154*	0.058**	0.036**	−0.025	−0.119
	(0.007)	(0.009)	(0.023)	(0.076)	(0.009)	(0.013)	(0.031)	(0.088)
2003	0.068**	0.026**	−0.026*	−0.099**	0.097**	0.052**	0.026	−0.014
	(0.007)	(0.008)	(0.013)	(0.031)	(0.010)	(0.013)	(0.023)	(0.073)
2004	0.115**	0.030**	−0.067**	−0.134**	0.134**	0.087**	0.018	−0.005
	(0.008)	(0.007)	(0.008)	(0.018)	(0.011)	(0.011)	(0.012)	(0.035)
2005	0.168**	0.070**	−0.079**	−0.108**	0.128**	0.089**	−0.044**	−0.076**
	(0.010)	(0.009)	(0.008)	(0.012)	(0.010)	(0.010)	(0.010)	(0.018)
2006	0.152**	−0.044**	−0.140**	−0.231**	0.094**	0.049**	−0.035**	−0.243**
	(0.012)	(0.010)	(0.008)	(0.009)	(0.011)	(0.011)	(0.010)	(0.014)
2007	0.244**	0.013	−0.138**	−0.199**	0.147**	0.062**	−0.018	−0.083**
	(0.013)	(0.012)	(0.010)	(0.013)	(0.013)	(0.013)	(0.013)	(0.018)

NOTE: Models include controls for nonhostile deployment only in window, years of service at the time of the decision, education, gender, AFQT, race, and an indicator for being promoted more rapidly than is typical.
** = statistically significant at the 1-percent level. * = statistically significant at the 5-percent level.

Table C.36
Effect of Months of Deployment in Prior Three Years on Reenlistment, by Year of Decision, Navy

Year of Decision	First Term				Second Term			
	1–6 Months	7–11 Months	12–17 Months	18+ Months	1–6 Months	7–11 Months	12–17 Months	18+ Months
1996	0.006	0.009	0.080	−0.261**	0.092**	0.103**	0.125	0.098
	(0.007)	(0.015)	(0.201)	(0.076)	(0.006)	(0.019)	(0.082)	(0.118)
1997	0.019**	−0.006	−0.072	0.000**	0.120**	0.125**	0.128	0.037
	(0.006)	(0.014)	(0.110)	(0.000)	(0.007)	(0.021)	(0.090)	(0.160)
1998	0.012	−0.004	−0.159*	−0.221**	0.131**	0.168**	0.142†	0.175†
	(0.008)	(0.014)	(0.072)	(0.092)	(0.008)	(0.019)	(0.079)	(0.096)
1999	−0.008	−0.031**	−0.105†	−0.034	0.112**	0.135**	0.126†	0.210**
	(0.009)	(0.013)	(0.056)	(0.083)	(0.009)	(0.017)	(0.070)	(0.061)
2000	−0.001	−0.024†	0.039	0.074	0.088**	0.088**	0.174**	0.094
	(0.009)	(0.012)	(0.053)	(0.083)	(0.009)	(0.017)	(0.060)	(0.061)
2001	−0.005	−0.052**	−0.080†	−0.073	0.083**	0.015	0.144**	0.105*
	(0.009)	(0.014)	(0.045)	(0.086)	(0.009)	(0.017)	(0.052)	(0.052)
2002	0.003	−0.077**	−0.056	−0.021	0.072**	0.041*	0.103**	0.087
	(0.007)	(0.012)	(0.046)	(0.051)	(0.009)	(0.018)	(0.043)	(0.056)
2003	0.027**	−0.021†	0.084*	−0.027	0.039**	−0.023	0.061	0.100†
	(0.009)	(0.012)	(0.041)	(0.087)	(0.010)	(0.018)	(0.051)	(0.054)
2004	0.029**	0.015	−0.022	0.095	0.056**	0.038**	0.017	0.073
	(0.008)	(0.010)	(0.033)	(0.061)	(0.010)	(0.015)	(0.045)	(0.049)
2005	0.007	−0.006	−0.018	−0.038	0.077**	0.079**	0.123**	0.041
	(0.008)	(0.010)	(0.028)	(0.052)	(0.010)	(0.014)	(0.036)	(0.048)
2006	0.040**	0.012	−0.053†	−0.029	0.087**	0.095**	0.120**	0.151**
	(0.008)	(0.012)	(0.027)	(0.057)	(0.009)	(0.015)	(0.035)	(0.044)
2007	0.013	0.007	−0.024	0.014	0.094**	0.111**	0.109**	0.025
	(0.010)	(0.014)	(0.026)	(0.051)	(0.011)	(0.018)	(0.037)	(0.059)

NOTE: Models include controls for nonhostile deployment only in window, years of service at the time of the decision, education, gender, AFQT, race, and an indicator for being promoted more rapidly than is typical.
** = statistically significant at the 1-percent level. * = statistically significant at the 5-percent level. † = statistically significant at the 10-percent level.

Table C.37
Effect of Months of Deployment in Prior Three Years on Reenlistment, by Year of Decision, Marine Corps

Year of Decision	First Term				Second Term			
	1–6 Months	7–11 Months	12–17 Months	18+ Months	1–6 Months	7–11 Months	12–17 Months	18+ Months
1996	0.016**	−0.051†	0.169**	−0.042	0.106**	0.162*	0.086	0.043
	(0.006)	(0.027)	(0.072)	(0.156)	(0.015)	(0.071)	(0.114)	(0.080)
1997	0.008	−0.028*	0.001	0.021	0.139**	0.225**	0.109	0.498**
	(0.007)	(0.013)	(0.046)	(0.091)	(0.019)	(0.069)	(0.109)	(0.125)
1998	0.009	−0.056**	0.036	0.280†	0.135**	0.205**	0.147	0.000**
	(0.007)	(0.012)	(0.044)	(0.165)	(0.022)	(0.060)	(0.129)	(0.000)
1999	0.001	−0.032**	0.001	0.063	0.117**	0.081	0.161	0.162
	(0.006)	(0.012)	(0.042)	(0.086)	(0.023)	(0.067)	(0.151)	(0.215)
2000	0.016**	−0.041**	0.021	−0.011	0.081**	0.006	0.325†	0.167
	(0.007)	(0.011)	(0.031)	(0.063)	(0.021)	(0.056)	(0.170)	(0.191)
2001	0.001	−0.111**	−0.026	0.176†	0.089**	0.123*	0.015	−0.390**
	(0.006)	(0.013)	(0.029)	(0.105)	(0.021)	(0.054)	(0.170)	(0.150)
2002	0.009	−0.055**	0.009	0.046	0.099**	0.043	−0.004	−0.417**
	(0.006)	(0.011)	(0.028)	(0.065)	(0.020)	(0.056)	(0.083)	(0.135)
2003	−0.001	−0.004	−0.021	0.047	0.046*	0.012	0.005	0.025
	(0.006)	(0.008)	(0.020)	(0.051)	(0.020)	(0.031)	(0.071)	(0.121)
2004	0.010	−0.037**	−0.078**	−0.055	0.074**	0.074**	0.061	−0.094
	(0.007)	(0.008)	(0.012)	(0.056)	(0.019)	(0.026)	(0.044)	(0.095)
2005	0.011	−0.025**	−0.089**	−0.065**	0.098**	0.074**	0.065**	0.036
	(0.008)	(0.008)	(0.009)	(0.019)	(0.019)	(0.021)	(0.026)	(0.052)
2006	0.048**	0.030**	−0.007	−0.031**	0.163**	0.102**	0.107**	0.031
	(0.009)	(0.008)	(0.008)	(0.011)	(0.019)	(0.020)	(0.024)	(0.051)
2007	0.118**	0.028**	−0.026**	−0.069**	0.128**	0.087**	0.029	0.090†
	(0.012)	(0.011)	(0.011)	(0.016)	(0.021)	(0.020)	(0.025)	(0.049)

NOTE: Models include controls for nonhostile deployment only in window, years of service at the time of the decision, education, gender, AFQT, race, and an indicator for being promoted more rapidly than is typical.
** = statistically significant at the 1-percent level. * = statistically significant at the 5-percent level. † = statistically significant at the 10-percent level.

Table C.38
Effect of Months of Deployment in Prior Three Years on Reenlistment, by Year of Decision, Air Force

Year of Decision	First Term				Second Term			
	1–6 Months	7–11 Months	12–17 Months	18+ Months	1–6 Months	7–11 Months	12–17 Months	18+ Months
1996	0.047**	−0.031†	0.045	0.036	0.061**	0.072**	0.104**	0.074**
	(0.008)	(0.017)	(0.030)	(0.047)	(0.006)	(0.012)	(0.018)	(0.025)
1997	0.028**	−0.030†	−0.048	0.000	0.058**	0.039**	0.120**	0.106**
	(0.008)	(0.018)	(0.031)	(0.049)	(0.007)	(0.016)	(0.023)	(0.029)
1998	0.000	−0.055**	−0.042	0.081	0.050**	0.039**	0.072**	0.049
	(0.008)	(0.015)	(0.032)	(0.052)	(0.008)	(0.017)	(0.025)	(0.032)
1999	0.010	−0.049**	0.024	0.047	0.054**	0.086**	0.087**	0.100**
	(0.008)	(0.016)	(0.031)	(0.051)	(0.009)	(0.018)	(0.028)	(0.031)
2000	0.009	−0.104**	−0.016	0.064	0.052**	0.064**	0.093**	0.055
	(0.008)	(0.015)	(0.031)	(0.051)	(0.008)	(0.017)	(0.027)	(0.036)
2001	0.002	−0.070**	−0.079*	−0.064	0.046**	0.025	0.062†	0.028
	(0.008)	(0.017)	(0.036)	(0.062)	(0.010)	(0.020)	(0.032)	(0.042)
2002	0.002	−0.025†	0.038	0.042	0.036**	0.024	0.060*	0.081*
	(0.008)	(0.015)	(0.035)	(0.052)	(0.009)	(0.019)	(0.030)	(0.035)
2003	0.007	−0.063**	0.026	−0.019	0.033**	−0.014	0.078**	0.074†
	(0.010)	(0.017)	(0.034)	(0.062)	(0.011)	(0.021)	(0.031)	(0.041)
2004	0.027**	−0.005	−0.119**	0.009	0.025*	0.008	0.042	−0.049
	(0.009)	(0.013)	(0.025)	(0.049)	(0.013)	(0.020)	(0.035)	(0.057)
2005	0.018*	−0.005	−0.051**	−0.068†	0.050**	0.031†	0.050†	0.036
	(0.008)	(0.012)	(0.020)	(0.039)	(0.013)	(0.018)	(0.030)	(0.053)
2006	0.029**	0.006	−0.013	−0.068†	0.033**	0.036*	0.004	−0.041
	(0.008)	(0.012)	(0.018)	(0.035)	(0.012)	(0.017)	(0.027)	(0.050)
2007	0.036**	−0.020	−0.027	−0.033	0.042**	0.055**	0.046†	−0.002
	(0.009)	(0.013)	(0.021)	(0.040)	(0.013)	(0.018)	(0.028)	(0.060)

NOTE: Models include controls for nonhostile deployment only in window, years of service at the time of the decision, education, gender, AFQT, race, and an indicator for being promoted more rapidly than is typical.
** = statistically significant at the 1-percent level. * = statistically significant at the 5-percent level. † = statistically significant at the 10-percent level.

Table C.39
Effect of Number of Deployments in Prior Three Years on Reenlistment, by Year of Decision, Army

Year of Decision	First Term			Second Term		
	1 Deployment	2 Deployments	3+ Deployments	1 Deployment	2 Deployments	3+ Deployments
1996	0.055**	0.060**	0.099**	0.088**	0.104**	0.122**
	(0.006)	(0.014)	(0.038)	(0.006)	(0.012)	(0.022)
1997	0.077**	0.126**	0.186**	0.091**	0.112**	0.132**
	(0.007)	(0.013)	(0.029)	(0.007)	(0.012)	(0.022)
1998	0.071**	0.093**	0.088**	0.080**	0.095**	0.075**
	(0.007)	(0.013)	(0.024)	(0.009)	(0.015)	(0.024)
1999	0.058**	0.076**	0.024	0.068**	0.080**	0.065**
	(0.007)	(0.012)	(0.024)	(0.008)	(0.013)	(0.023)
2000	0.059**	0.058**	0.032	0.068**	0.074**	−0.006
	(0.007)	(0.011)	(0.025)	(0.009)	(0.014)	(0.027)
2001	0.036**	0.050**	0.037	0.054**	0.065**	0.059*
	(0.006)	(0.009)	(0.023)	(0.008)	(0.012)	(0.026)
2002	0.030**	0.017†	−0.008	0.048**	0.047**	0.035
	(0.007)	(0.010)	(0.021)	(0.009)	(0.013)	(0.025)
2003	0.060**	0.020	−0.032†	0.085**	0.077**	0.013
	(0.007)	(0.009)	(0.018)	(0.010)	(0.013)	(0.027)
2004	0.037**	−0.005	0.049**	0.092**	0.069**	0.094**
	(0.007)	(0.008)	(0.013)	(0.010)	(0.012)	(0.017)
2005	−0.023**	0.111**	−0.003	0.019*	0.105**	0.060**
	(0.008)	(0.010)	(0.012)	(0.009)	(0.010)	(0.014)
2006	−0.068**	−0.115**	−0.140**	0.011	−0.023*	−0.013
	(0.008)	(0.009)	(0.015)	(0.010)	(0.011)	(0.017)
2007	−0.052**	−0.043**	−0.071**	0.031**	0.053**	0.093**
	(0.010)	(0.011)	(0.021)	(0.012)	(0.013)	(0.023)

NOTE: Models include controls for nonhostile deployment only in window, years of service at the time of the decision, education, gender, AFQT, race, and an indicator for being promoted more rapidly than is typical. ** = statistically significant at the 1-percent level. * = statistically significant at the 5-percent level. † = statistically significant at the 10-percent level.

Table C.40
Effect of Number of Deployments in Prior Three Years on Reenlistment, by Year of Decision, Navy

Year of Decision	First Term			Second Term		
	1 Deployment	2 Deployments	3+ Deployments	1 Deployment	2 Deployments	3+ Deployments
1996	0.003	0.007	0.039	0.088**	0.098**	0.134**
	(0.008)	(0.009)	(0.020)	(0.007)	(0.010)	(0.020)
1997	0.023**	0.001	0.018	0.115**	0.134**	0.130**
	(0.007)	(0.009)	(0.013)	(0.008)	(0.012)	(0.019)
1998	0.012	−0.004	0.048*	0.115**	0.170**	0.240**
	(0.008)	(0.010)	(0.021)	(0.009)	(0.013)	(0.023)
1999	−0.003	−0.025**	−0.030[†]	0.111**	0.121**	0.147**
	(0.009)	(0.011)	(0.016)	(0.010)	(0.013)	(0.021)
2000	0.004	−0.017[†]	−0.007	0.096**	0.067**	0.114**
	(0.009)	(0.010)	(0.017)	(0.009)	(0.012)	(0.020)
2001	−0.005	−0.020[†]	−0.021	0.085**	0.069**	0.039*
	(0.009)	(0.011)	(0.016)	(0.010)	(0.012)	(0.020)
2002	0.009	−0.034**	−0.039**	0.078**	0.054**	0.051**
	(0.008)	(0.009)	(0.015)	(0.010)	(0.012)	(0.022)
2003	0.038**	−0.011	0.019	0.049**	0.008	0.006
	(0.010)	(0.010)	(0.013)	(0.011)	(0.014)	(0.020)
2004	0.026**	0.017[†]	0.037**	0.052**	0.055**	0.046**
	(0.009)	(0.009)	(0.012)	(0.010)	(0.013)	(0.019)
2005	0.007	0.001	−0.007	0.077**	0.076**	0.088**
	(0.009)	(0.010)	(0.012)	(0.010)	(0.013)	(0.016)
2006	0.032**	0.031**	0.029*	0.090**	0.092**	0.086**
	(0.009)	(0.010)	(0.014)	(0.009)	(0.012)	(0.018)
2007	0.017[†]	−0.006	0.021	0.094**	0.096**	0.124**
	(0.010)	(0.012)	(0.019)	(0.011)	(0.016)	(0.026)

NOTE: Models include controls for nonhostile deployment only in window, years of service at the time of the decision, education, gender, AFQT, race, and an indicator for being promoted more rapidly than is typical.
** = statistically significant at the 1-percent level. * = statistically significant at the 5-percent level. [†] = statistically significant at the 10-percent level.

Table C.41
Effect of Number of Deployments in Prior Three Years on Reenlistment, by Year of Decision, Marine Corps

Year of Decision	First Term			Second Term		
	1 Deployment	2 Deployments	3+ Deployments	1 Deployment	2 Deployments	3+ Deployments
1996	0.023**	−0.003	0.007	0.108**	0.090**	0.137**
	(0.007)	(0.010)	(0.024)	(0.016)	(0.030)	(0.058)
1997	0.009	−0.012	0.020	0.146**	0.126**	0.170†
	(0.007)	(0.010)	(0.033)	(0.020)	(0.040)	(0.092)
1998	0.018*	−0.033**	−0.053**	0.128**	0.192**	0.147*
	(0.008)	(0.010)	(0.022)	(0.023)	(0.042)	(0.070)
1999	0.004	−0.011	−0.034*	0.113**	0.092**	0.233**
	(0.007)	(0.009)	(0.016)	(0.025)	(0.038)	(0.068)
2000	0.017*	−0.017†	0.007	0.079**	0.094**	0.023
	(0.007)	(0.009)	(0.015)	(0.022)	(0.038)	(0.077)
2001	0.007	−0.043**	−0.069**	0.082**	0.103**	0.091
	(0.007)	(0.010)	(0.018)	(0.022)	(0.038)	(0.069)
2002	0.009	−0.031**	0.036†	0.092**	0.090**	0.056
	(0.007)	(0.010)	(0.019)	(0.021)	(0.035)	(0.073)
2003	0.008	−0.014†	−0.040**	0.037†	0.036	0.042
	(0.006)	(0.008)	(0.011)	(0.021)	(0.028)	(0.052)
2004	−0.004	−0.024**	−0.031**	0.070**	0.074**	0.076†
	(0.007)	(0.008)	(0.011)	(0.019)	(0.025)	(0.041)
2005	−0.010	−0.048**	−0.049**	0.074**	0.101**	0.073**
	(0.008)	(0.008)	(0.010)	(0.018)	(0.021)	(0.030)
2006	0.034**	0.019*	−0.031**	0.130**	0.127**	0.109**
	(0.008)	(0.008)	(0.010)	(0.018)	(0.020)	(0.031)
2007	0.030**	0.028**	0.012	0.094**	0.069**	0.155**
	(0.010)	(0.011)	(0.014)	(0.018)	(0.023)	(0.038)

NOTE: Models include controls for nonhostile deployment only in window, years of service at the time of the decision, education, gender, AFQT, race, and an indicator for being promoted more rapidly than is typical.
** = statistically significant at the 1-percent level. * = statistically significant at the 5-percent level. † = statistically significant at the 10-percent level.

Table C.42
Effect of Number of Deployments in Prior Three Years on Reenlistment, by Year of Decision, Air Force

Year of Decision	First Term			Second Term		
	1 Deployment	**2 Deployments**	**3+ Deployments**	**1 Deployment**	**2 Deployments**	**3+ Deployments**
1996	0.037**	0.034**	0.044*	0.053**	0.094**	0.088**
	(0.008)	(0.013)	(0.020)	(0.006)	(0.010)	(0.012)
1997	0.030**	0.005	−0.031†	0.068**	0.043**	0.041**
	(0.009)	(0.014)	(0.018)	(0.008)	(0.012)	(0.015)
1998	0.011	−0.015	−0.088**	0.051**	0.049**	0.047**
	(0.009)	(0.013)	(0.016)	(0.008)	(0.013)	(0.016)
1999	0.015†	−0.005	−0.047**	0.062**	0.065**	0.051**
	(0.009)	(0.013)	(0.017)	(0.009)	(0.014)	(0.017)
2000	0.015†	−0.018	−0.106**	0.049**	0.077**	0.051**
	(0.008)	(0.012)	(0.016)	(0.009)	(0.013)	(0.017)
2001	0.009	−0.037**	−0.093**	0.049**	0.032*	0.033
	(0.009)	(0.014)	(0.021)	(0.010)	(0.015)	(0.021)
2002	0.006	−0.013	−0.027	0.042**	0.018	0.035†
	(0.008)	(0.013)	(0.019)	(0.010)	(0.015)	(0.021)
2003	0.016	−0.038**	−0.062**	0.035**	0.038*	−0.026
	(0.010)	(0.015)	(0.021)	(0.012)	(0.017)	(0.024)
2004	0.022*	0.008	−0.021	0.022†	0.023	0.014
	(0.010)	(0.012)	(0.017)	(0.013)	(0.018)	(0.022)
2005	0.011	−0.010	0.018	0.056**	0.021	0.041†
	(0.009)	(0.012)	(0.015)	(0.013)	(0.018)	(0.021)
2006	0.018*	0.014	0.040**	0.033**	0.016	0.044*
	(0.008)	(0.011)	(0.016)	(0.012)	(0.017)	(0.021)
2007	0.027**	−0.006	0.008	0.039**	0.060**	0.046†
	(0.009)	(0.013)	(0.019)	(0.013)	(0.018)	(0.024)

NOTE: Models include controls for nonhostile deployment only in window, years of service at the time of the decision, education, gender, AFQT, race, and an indicator for being promoted more rapidly than is typical.
** = statistically significant at the 1-percent level. * = statistically significant at the 5-percent level. † = statistically significant at the 10-percent level.

Table C.43
Effect of Number of Deployments in Prior Year on Reenlistment, by Gender and Year of Decision, First Term

Year of Decision	Army		Navy		Marine Corps		Air Force	
	Female	Male	Female	Male	Female	Male	Female	Male
1996	0.105**	0.052**	0.055	0.014†	0.472**	0.017†	0.057**	0.026**
	(0.030)	(0.008)	(0.039)	(0.007)	(0.152)	(0.009)	(0.022)	(0.009)
1997	0.132**	0.126**	0.030	0.009	0.213	−0.007	0.035	0.014
	(0.022)	(0.008)	(0.036)	(0.006)	(0.161)	(0.008)	(0.023)	(0.009)
1998	0.085**	0.061**	0.015	−0.011	0.168	−0.014†	0.033	−0.016†
	(0.025)	(0.009)	(0.026)	(0.008)	(0.172)	(0.008)	(0.021)	(0.009)
1999	0.071**	0.053**	0.032	−0.017*	−0.039	0.004	0.004	0.000
	(0.022)	(0.008)	(0.026)	(0.008)	(0.076)	(0.007)	(0.023)	(0.009)
2000	0.073**	0.079**	0.040†	−0.014†	0.173†	−0.000	0.034†	−0.011
	(0.021)	(0.008)	(0.023)	(0.008)	(0.095)	(0.007)	(0.020)	(0.009)
2001	0.064**	0.046**	0.030	−0.067**	0.044	−0.032**	−0.006	−0.016
	(0.024)	(0.007)	(0.024)	(0.008)	(0.074)	(0.008)	(0.021)	(0.010)
2002	0.025	0.024**	0.056**	−0.045**	0.012	0.000	0.015	−0.015
	(0.024)	(0.008)	(0.019)	(0.007)	(0.043)	(0.007)	(0.019)	(0.009)
2003	0.015	0.047**	0.076**	−0.022**	0.001	0.006	−0.005	−0.029**
	(0.017)	(0.007)	(0.018)	(0.008)	(0.025)	(0.006)	(0.019)	(0.011)
2004	0.028*	0.022**	0.059**	0.006	0.045†	−0.002	0.003	−0.023*
	(0.014)	(0.007)	(0.017)	(0.007)	(0.027)	(0.006)	(0.018)	(0.010)
2005	0.036	0.015**	0.017	−0.018**	0.033	0.002	0.010	−0.004
	(0.015)	(0.006)	(0.018)	(0.008)	(0.032)	(0.006)	(0.017)	(0.009)
2006	−0.053**	−0.085**	0.077**	0.005	0.067**	0.021**	−0.010	0.023**
	(0.016)	(0.006)	(0.020)	(0.008)	(0.024)	(0.005)	(0.018)	(0.009)
2007	−0.008	−0.058**	0.042†	−0.015†	0.088**	0.062**	0.016	0.025**
	(0.020)	(0.008)	(0.023)	(0.009)	(0.032)	(0.007)	(0.020)	(0.010)

NOTE: Models include controls for nonhostile deployment only in window, years of service at the time of the decision, education, gender, AFQT, race, and an indicator for being promoted more rapidly than is typical.
** = statistically significant at the 1-percent level. * = statistically significant at the 5-percent level. † = statistically significant at the 10-percent level.

Table C.44
Effect of Number of Deployments in Prior Year on Reenlistment, by Gender and Year of Decision, Second Term

Year of Decision	Army		Navy		Marine Corps		Air Force	
	Female	Male	Female	Male	Female	Male	Female	Male
1996	0.075**	0.093**	0.146**	0.083**	−0.546**	0.105**	0.111**	0.059**
	(0.028)	(0.007)	(0.036)	(0.008)	(0.147)	(0.021)	(0.021)	(0.007)
1997	0.129**	0.104**	0.044	0.086**	0.000**	0.150**	0.095**	0.055**
	(0.026)	(0.008)	(0.046)	(0.009)	(0.000)	(0.024)	(0.026)	(0.008)
1998	0.119**	0.089**	0.176**	0.130**	−0.387**	0.166**	0.028	0.054**
	(0.033)	(0.010)	(0.040)	(0.009)	(0.124)	(0.027)	(0.027)	(0.009)
1999	0.090**	0.104**	0.165**	0.113**	−1.319**	0.129**	0.104**	0.060**
	(0.032)	(0.010)	(0.043)	(0.010)	(0.414)	(0.027)	(0.027)	(0.010)
2000	0.117**	0.103**	0.175**	0.050**	−0.169	0.093**	0.066**	0.069**
	(0.031)	(0.010)	(0.031)	(0.010)	(0.426)	(0.025)	(0.024)	(0.009)
2001	0.093**	0.092**	0.063[†]	0.030**	−0.494[†]	0.091**	0.069**	0.055**
	(0.031)	(0.010)	(0.034)	(0.011)	(0.288)	(0.029)	(0.029)	(0.012)
2002	0.026	0.058**	0.103**	0.009	0.027	0.068**	0.052*	0.020[†]
	(0.034)	(0.011)	(0.031)	(0.010)	(0.196)	(0.024)	(0.026)	(0.011)
2003	0.086**	0.070**	0.009	−0.006	−0.098	0.015	−0.003	0.014
	(0.025)	(0.010)	(0.034)	(0.011)	(0.131)	(0.019)	(0.029)	(0.012)
2004	0.096**	0.078**	0.072**	0.018[†]	0.277*	0.049**	0.039	0.004
	(0.023)	(0.009)	(0.030)	(0.011)	(0.129)	(0.017)	(0.028)	(0.014)
2005	0.096**	−0.011	0.133**	0.060**	0.181	0.064**	0.052[†]	0.031*
	(0.018)	(0.007)	(0.030)	(0.010)	(0.112)	(0.016)	(0.030)	(0.014)
2006	−0.010	−0.093**	0.107**	0.060**	0.051	0.109**	0.051[†]	0.044**
	(0.020)	(0.007)	(0.027)	(0.010)	(0.084)	(0.015)	(0.029)	(0.013)
2007	0.042[†]	−0.006	0.131**	0.037**	0.231**	0.091**	0.053[†]	0.033**
	(0.024)	(0.009)	(0.033)	(0.012)	(0.098)	(0.016)	(0.029)	(0.014)

NOTE: Models include controls for nonhostile deployment only in window, years of service at the time of the decision, education, gender, AFQT, race, and an indicator for being promoted more rapidly than is typical.
** = statistically significant at the 1-percent level. * = statistically significant at the 5-percent level. [†] = statistically significant at the 10-percent level.

Table C.45
Effect of Number of Deployments in Prior Year on Reenlistment, by Combat Arms and Year of Decision, First Term

Year of Decision	Army		Navy		Marine Corps		Air Force	
	Non-Combat Arms	Combat Arms	Non-Combat Arms	Combat Arms	Non-Combat Arms	Combat Arms	Non-Combat Arms	Combat Arms
1996	0.067**	0.032*	0.018*	0.003	0.018	0.020†	0.031**	0.011
	(0.010)	(0.014)	(0.008)	(0.016)	(0.013)	(0.012)	(0.009)	(0.025)
1997	0.110**	0.167**	0.006	0.027*	−0.010	−0.001	0.027**	−0.062**
	(0.009)	(0.014)	(0.007)	(0.012)	(0.011)	(0.010)	(0.009)	(0.021)
1998	0.060**	0.063**	−0.007	−0.004	−0.013	−0.014	−0.002	−0.046**
	(0.010)	(0.013)	(0.008)	(0.015)	(0.011)	(0.011)	(0.009)	(0.017)
1999	0.050**	0.064**	−0.015†	0.017	0.012	−0.005	0.008	−0.036
	(0.009)	(0.013)	(0.008)	(0.018)	(0.009)	(0.010)	(0.009)	(0.023)
2000	0.088**	0.062**	−0.016*	0.071**	0.000	0.008	0.001	−0.040†
	(0.009)	(0.013)	(0.008)	(0.021)	(0.010)	(0.011)	(0.008)	(0.024)
2001	0.056**	0.040**	−0.062**	0.002	−0.009	−0.053**	−0.012	−0.034
	(0.009)	(0.011)	(0.008)	(0.024)	(0.010)	(0.012)	(0.010)	(0.023)
2002	0.032**	0.007	−0.032**	−0.009	0.016†	−0.013	−0.014	0.021
	(0.009)	(0.012)	(0.007)	(0.025)	(0.009)	(0.011)	(0.009)	(0.021)
2003	0.053**	0.021†	−0.006	0.007	−0.004	0.033**	−0.021*	−0.044
	(0.007)	(0.011)	(0.008)	(0.023)	(0.006)	(0.011)	(0.010)	(0.028)
2004	0.028**	0.000	0.015*	0.021	−0.000	−0.003	−0.013	−0.033
	(0.007)	(0.012)	(0.007)	(0.018)	(0.007)	(0.013)	(0.009)	(0.024)
2005	0.034**	−0.025**	−0.013†	−0.015	−0.001	0.019	−0.003	−0.005
	(0.007)	(0.010)	(0.007)	(0.020)	(0.006)	(0.014)	(0.009)	(0.020)
2006	−0.064**	−0.129**	0.019**	−0.026	0.029**	0.002	0.018*	−0.000
	(0.007)	(0.012)	(0.008)	(0.022)	(0.006)	(0.011)	(0.009)	(0.019)
2007	−0.036**	−0.089**	−0.012	0.038	0.057**	0.086**	0.025**	0.007
	(0.009)	(0.015)	(0.008)	(0.028)	(0.008)	(0.014)	(0.009)	(0.025)

NOTE: Models include controls for nonhostile deployment only in window, years of service at the time of the decision, education, gender, AFQT, race, and an indicator for being promoted more rapidly than is typical.
** = statistically significant at the 1-percent level. * = statistically significant at the 5-percent level. † = statistically significant at the 10-percent level.

Table C.46
Effect of Number of Deployments in Prior Year on Reenlistment, by Combat Arms and Year of Decision, Second Term

Year of Decision	Army		Navy		Marine Corps		Air Force	
	Non-Combat Arms	Combat Arms	Non-Combat Arms	Combat Arms	Non-Combat Arms	Combat Arms	Non-Combat Arms	Combat Arms
1996	0.098**	0.080**	0.079**	0.175**	0.099**	0.117**	0.067**	0.044*
	(0.009)	(0.013)	(0.008)	(0.025)	(0.025)	(0.037)	(0.007)	(0.021)
1997	0.110**	0.097**	0.077**	0.198**	0.143**	0.159**	0.065**	−0.000
	(0.009)	(0.013)	(0.009)	(0.030)	(0.029)	(0.044)	(0.008)	(0.025)
1998	0.088**	0.096**	0.121**	0.241**	0.124**	0.263**	0.058**	0.003
	(0.011)	(0.017)	(0.010)	(0.029)	(0.032)	(0.048)	(0.009)	(0.024)
1999	0.111**	0.084**	0.111**	0.176**	0.114**	0.158**	0.068**	0.042
	(0.011)	(0.018)	(0.010)	(0.032)	(0.033)	(0.049)	(0.010)	(0.029)
2000	0.111**	0.087**	0.056**	0.118**	0.053†	0.182**	0.076**	0.026
	(0.011)	(0.017)	(0.009)	(0.032)	(0.029)	(0.050)	(0.009)	(0.028)
2001	0.088**	0.099**	0.024*	0.130**	0.068*	0.132**	0.056**	0.072*
	(0.012)	(0.017)	(0.011)	(0.036)	(0.034)	(0.053)	(0.011)	(0.032)
2002	0.068**	0.026	0.011	0.107**	0.057*	0.085†	0.026**	0.014
	(0.012)	(0.018)	(0.010)	(0.035)	(0.028)	(0.045)	(0.011)	(0.026)
2003	0.086**	0.035†	−0.012	0.110**	0.007	0.025	0.011	0.009
	(0.010)	(0.019)	(0.011)	(0.037)	(0.021)	(0.040)	(0.012)	(0.037)
2004	0.097**	0.034*	0.022*	0.101**	0.058**	0.046	0.005	0.074†
	(0.010)	(0.017)	(0.010)	(0.033)	(0.019)	(0.037)	(0.013)	(0.038)
2005	0.040**	−0.098**	0.061**	0.173**	0.096**	−0.034	0.036**	0.054
	(0.008)	(0.013)	(0.010)	(0.032)	(0.018)	(0.034)	(0.013)	(0.040)
2006	−0.063**	−0.139**	0.062**	0.091**	0.121**	0.043	0.042**	0.061†
	(0.008)	(0.013)	(0.009)	(0.033)	(0.017)	(0.032)	(0.012)	(0.036)
2007	0.020*	−0.066**	0.046**	0.118**	0.107**	0.078*	0.039**	0.033
	(0.009)	(0.018)	(0.011)	(0.044)	(0.018)	(0.036)	(0.013)	(0.037)

NOTE: Models include controls for nonhostile deployment only in window, years of service at the time of the decision, education, gender, AFQT, race, and an indicator for being promoted more rapidly than is typical.
** = statistically significant at the 1-percent level. * = statistically significant at the 5-percent level. † = statistically significant at the 10-percent level.

Table C.47
Effect of Number of Deployments in Prior Year on Reenlistment, by Marital Status and Year of Decision, First Term

Year of Decision	Army		Navy		Marine Corps		Air Force	
	Married	Single	Married	Single	Married	Single	Married	Single
1996	0.088**	0.035**	0.035**	−0.005	0.048**	−0.007	0.049**	0.028**
	(0.014)	(0.009)	(0.013)	(0.008)	(0.016)	(0.010)	(0.012)	(0.011)
1997	0.145**	0.121**	0.035**	−0.015*	0.011	−0.022**	0.036**	0.008
	(0.013)	(0.009)	(0.012)	(0.007)	(0.015)	(0.008)	(0.013)	(0.011)
1998	0.124**	0.039**	0.010	−0.025**	0.014	−0.028**	−0.002	−0.002
	(0.016)	(0.009)	(0.014)	(0.009)	(0.016)	(0.009)	(0.012)	(0.010)
1999	0.096**	0.042**	0.018	−0.030**	0.020	−0.010	0.027*	−0.008
	(0.014)	(0.009)	(0.015)	(0.009)	(0.013)	(0.008)	(0.013)	(0.011)
2000	0.133**	0.056**	0.032*	−0.024**	0.011	−0.007	0.027*	−0.014
	(0.013)	(0.009)	(0.015)	(0.008)	(0.014)	(0.008)	(0.012)	(0.011)
2001	0.073**	0.038**	−0.008	−0.068**	−0.005	−0.045**	0.011	−0.022†
	(0.014)	(0.008)	(0.018)	(0.009)	(0.015)	(0.009)	(0.014)	(0.012)
2002	0.051**	0.013	0.033*	−0.042**	0.023†	−0.013†	0.029*	−0.029**
	(0.014)	(0.008)	(0.015)	(0.008)	(0.013)	(0.008)	(0.013)	(0.011)
2003	0.033**	0.045**	0.027†	−0.020	0.016	−0.015*	0.003	−0.028*
	(0.012)	(0.007)	(0.014)	(0.008)	(0.010)	(0.007)	(0.015)	(0.013)
2004	0.018†	0.026**	0.034**	−0.003	0.022*	−0.024**	0.019	−0.032**
	(0.011)	(0.007)	(0.012)	(0.008)	(0.010)	(0.007)	(0.013)	(0.012)
2005	0.020*	0.022**	0.007	−0.043**	0.024**	−0.011	0.014	−0.008
	(0.010)	(0.007)	(0.012)	(0.009)	(0.010)	(0.007)	(0.012)	(0.011)
2006	−0.100**	−0.059**	0.005	0.005	0.036**	0.007	0.052**	−0.007
	(0.010)	(0.007)	(0.012)	(0.009)	(0.009)	(0.007)	(0.011)	(0.010)
2007	−0.039**	−0.050**	−0.007	−0.020*	0.076**	0.043**	0.045**	0.005
	(0.012)	(0.010)	(0.013)	(0.010)	(0.012)	(0.009)	(0.013)	(0.012)

NOTE: Models include controls for nonhostile deployment only in window, years of service at the time of the decision, education, gender, AFQT, race, and an indicator for being promoted more rapidly than is typical.
** = statistically significant at the 1-percent level. * = statistically significant at the 5-percent level. † = statistically significant at the 10-percent level.

Table C.48
Effect of Number of Deployments in Prior Year on Reenlistment, by Marital Status and Year of Decision, Second Term

Year of Decision	Army		Navy		Marine Corps		Air Force	
	Married	Single	Married	Single	Married	Single	Married	Single
1996	0.097**	0.076**	0.095**	0.067**	0.128**	0.031	0.069**	0.052**
	(0.008)	(0.016)	(0.009)	(0.014)	(0.023)	(0.052)	(0.007)	(0.014)
1997	0.115**	0.094**	0.120**	−0.004	0.169**	0.059	0.069**	0.038*
	(0.008)	(0.014)	(0.010)	(0.016)	(0.027)	(0.068)	(0.009)	(0.017)
1998	0.103**	0.076**	0.173**	0.044**	0.157**	0.177**	0.056**	0.040*
	(0.011)	(0.017)	(0.011)	(0.016)	(0.033)	(0.058)	(0.009)	(0.018)
1999	0.112**	0.085**	0.142**	0.072**	0.141**	0.116*	0.072**	0.047**
	(0.011)	(0.017)	(0.011)	(0.016)	(0.033)	(0.058)	(0.011)	(0.018)
2000	0.125**	0.075**	0.121**	0.001	0.111**	0.044	0.071**	0.073**
	(0.011)	(0.016)	(0.012)	(0.014)	(0.030)	(0.055)	(0.010)	(0.017)
2001	0.116**	0.065**	0.104**	−0.018	0.140**	−0.006	0.071**	0.043*
	(0.013)	(0.015)	(0.014)	(0.015)	(0.036)	(0.057)	(0.013)	(0.021)
2002	0.076**	0.025	0.080**	−0.024[†]	0.063*	0.107*	0.036**	0.015
	(0.013)	(0.016)	(0.013)	(0.015)	(0.029)	(0.047)	(0.012)	(0.019)
2003	0.080**	0.063**	0.062**	−0.035**	0.039[†]	−0.012	0.029*	−0.024
	(0.012)	(0.014)	(0.014)	(0.015)	(0.023)	(0.038)	(0.013)	(0.021)
2004	0.103**	0.047**	0.076**	−0.015	0.065**	0.039	0.014	0.032
	(0.011)	(0.014)	(0.013)	(0.015)	(0.021)	(0.035)	(0.015)	(0.023)
2005	0.017*	−0.012	0.083**	0.054**	0.084**	0.002	0.053**	0.015
	(0.008)	(0.011)	(0.013)	(0.015)	(0.019)	(0.033)	(0.015)	(0.024)
2006	−0.067**	−0.101**	0.076**	0.045**	0.134**	0.054[†]	0.054**	0.025
	(0.009)	(0.011)	(0.012)	(0.014)	(0.018)	(0.030)	(0.013)	(0.022)
2007	0.009	−0.020	0.063**	0.036*	0.128**	0.054	0.046**	0.027
	(0.010)	(0.015)	(0.015)	(0.017)	(0.019)	(0.035)	(0.015)	(0.023)

NOTE: Models include controls for nonhostile deployment only in window, years of service at the time of the decision, education, gender, AFQT, race, and an indicator for being promoted more rapidly than is typical.
** = statistically significant at the 1-percent level. * = statistically significant at the 5-percent level. [†] = statistically significant at the 10-percent level.

Table C.49
Effect of Reenlistment Bonus Multiplier and Deployment on Reenlistment, 1996–2007

Year of Decision	First Term				Second Term			
	Army	Navy	Marine Corps	Air Force	Army	Navy	Marine Corps	Air Force
SRB multiplier	0.032**	0.057**	0.062**	0.017**	0.037**	0.019**	0.020**	0.016**
	(0.002)	(0.002)	(0.001)	(0.001)	(0.003)	(0.002)	(0.004)	(0.002)
Any HFP deployment × year of decision								
1996	0.057**	0.008	0.034**	0.022*	0.099**	0.075**	0.124**	0.046**
	(0.008)	(0.007)	(0.010)	(0.009)	(0.008)	(0.008)	(0.024)	(0.008)
1997	0.117**	−0.002	0.003	0.018*	0.090**	0.064**	0.109**	0.044**
	(0.007)	(0.006)	(0.008)	(0.009)	(0.008)	(0.009)	(0.027)	(0.009)
1998	0.058**	−0.021**	0.018*	−0.006	0.081**	0.104**	0.136**	0.040**
	(0.008)	(0.007)	(0.008)	(0.008)	(0.010)	(0.010)	(0.028)	(0.009)
1999	0.058**	−0.020**	0.007	0.018*	0.099**	0.093**	0.131**	0.073**
	(0.007)	(0.007)	(0.007)	(0.008)	(0.010)	(0.009)	(0.027)	(0.010)
2000	0.081**	−0.011†	0.016*	0.006	0.101**	0.059**	0.100**	0.077**
	(0.007)	(0.007)	(0.007)	(0.008)	(0.009)	(0.009)	(0.025)	(0.009)
2001	0.068**	−0.051**	−0.015†	−0.009	0.082**	0.027**	0.083**	0.057**
	(0.007)	(0.007)	(0.008)	(0.009)	(0.010)	(0.010)	(0.028)	(0.011)
2002	0.041**	−0.047**	0.023**	−0.021*	0.060**	0.021*	0.063**	0.027**
	(0.008)	(0.007)	(0.007)	(0.009)	(0.010)	(0.010)	(0.023)	(0.010)
2003	0.021**	−0.010	−0.003	−0.031**	0.091**	−0.009	−0.011	0.009
	(0.006)	(0.007)	(0.006)	(0.009)	(0.008)	(0.010)	(0.017)	(0.010)
2004	0.040**	0.015*	−0.045**	−0.026**	0.092**	0.025**	0.052**	0.032**
	(0.005)	(0.006)	(0.006)	(0.009)	(0.007)	(0.010)	(0.016)	(0.011)
2005	0.021**	−0.006	0.017**	−0.019*	0.013*	0.070**	0.075**	0.037**
	(0.005)	(0.007)	(0.006)	(0.008)	(0.006)	(0.009)	(0.014)	(0.011)
2006	−0.083**	0.037**	0.019**	−0.002	−0.086**	0.075**	0.103**	0.032**
	(0.005)	(0.007)	(0.006)	(0.008)	(0.007)	(0.009)	(0.014)	(0.010)
2007	−0.043**	0.003	0.048**	0.024**	−0.004	0.047**	0.095**	0.026*
	(0.007)	(0.008)	(0.007)	(0.009)	(0.008)	(0.010)	(0.014)	(0.011)

NOTE: Models include controls for 3-digit MOS, nonhostile deployment only in window, years of service at the time of the decision, education, gender, AFQT, race, an indicator for being promoted more rapidly than is typical, and year-of-decision indicators. ** = statistically significant at the 1-percent level. * = statistically significant at the 5-percent level. † = statistically significant at the 10-percent level.

Table C.50
Descriptive Statistics for Administrative Data Sample, 1996–1997

Statistic	Army		Navy		Marine Corps		Air Force	
	First Term	Second Term	First Term	Second Term	First Term	Second Term	First Term	Second Term
Reenlisted	0.349	0.636	0.435	0.572	0.245	0.581	0.534	0.683
Nonhostile deployment only, prior 12 months	0.055	0.118	0.112	0.127	0.132	0.174	0.041	0.085
Hostile deployment, prior 12 months	0.342	0.242	0.359	0.193	0.331	0.166	0.248	0.196
Nonhostile deployment only, prior 36 months	0.103	0.205	0.079	0.176	0.156	0.338	0.075	0.159
Hostile deployment, prior 36 months	0.472	0.393	0.638	0.398	0.477	0.298	0.402	0.358
SRB multiplier	0.742	0.676	1.534	1.816	0.863	0.526	1.300	0.952
	(0.683)	(0.683)	(0.683)	(0.683)	(0.683)	(0.683)	(0.683)	(0.683)
Years of service	3.998	8.287	4.228	9.574	4.217	9.582	4.336	10.852
HS dropout or education missing	0.009	0.008	0.019	0.018	0.002	0.001	0.000	0.000
GED	0.066	0.057	0.053	0.043	0.033	0.045	0.001	0.000
At least some college	0.122	0.137	0.074	0.090	0.047	0.089	0.378	0.678
Male	0.849	0.861	0.839	0.876	0.945	0.938	0.755	0.819
AFQT below category IIIB or missing	0.016	0.070	0.003	0.059	0.006	0.036	0.002	0.022
AFQT category IIIB	0.271	0.331	0.364	0.274	0.316	0.310	0.200	0.215
AFQT category II	0.368	0.297	0.331	0.389	0.363	0.348	0.447	0.427
AFQT category I	0.052	0.036	0.031	0.071	0.037	0.036	0.056	0.055
White	0.636	0.543	0.601	0.630	0.706	0.627	0.736	0.742
Black	0.207	0.318	0.195	0.203	0.114	0.211	0.156	0.176
Promoted rapidly	0.643	0.594	0.443	0.566	0.786	0.470	0.854	0.658
Combat arms	0.271	0.252	0.105	0.080	0.283	0.186	0.109	0.082

Table C.50—Continued

Statistic	Army		Navy		Marine Corps		Air Force	
	First Term	Second Term	First Term	Second Term	First Term	Second Term	First Term	Second Term
Married	0.377	0.681	0.380	0.659	0.449	0.761	0.491	0.748
Decision in 1996	0.102	0.164	0.087	0.153	0.080	0.146	0.100	0.197
Decision in 1997	0.087	0.130	0.106	0.123	0.084	0.108	0.090	0.150
Decision in 1998	0.068	0.082	0.081	0.099	0.078	0.083	0.084	0.126
Decision in 1999	0.081	0.087	0.077	0.084	0.077	0.067	0.082	0.096
Decision in 2000	0.081	0.078	0.083	0.087	0.082	0.069	0.093	0.097
Decision in 2001	0.083	0.080	0.080	0.076	0.087	0.070	0.080	0.063
Decision in 2002	0.076	0.065	0.085	0.068	0.087	0.072	0.078	0.055
Decision in 2003	0.077	0.055	0.077	0.057	0.084	0.068	0.062	0.044
Decision in 2004	0.099	0.063	0.088	0.062	0.086	0.072	0.072	0.038
Decision in 2005	0.098	0.079	0.091	0.065	0.088	0.084	0.088	0.041
Decision in 2006	0.087	0.069	0.082	0.074	0.094	0.089	0.097	0.049
Decision in 2007	0.061	0.048	0.062	0.051	0.072	0.072	0.075	0.043
Number of observations	315,779	230,821	228,687	178,047	226,835	53,789	213,660	134,909

NOTE: Standard deviations of nonbinary variables are in parentheses.

Table C.51
Descriptive Statistics for Survey Sample, First-Term Respondents

Statistic	Army	Navy	Marine Corps	Air Force
High work stress	0.586	0.576	0.567	0.468
High personal stress	0.497	0.439	0.477	0.354
Intention to stay	0.346	0.453	0.304	0.473
Observe reenlistment decision	0.594	0.649	0.759	0.639
Reenlisted (conditional on reenlistment decision being observed)	0.394	0.419	0.270	0.512
Nonhostile deployment only	0.126	0.102	0.111	0.067
Hostile deployment	0.296	0.290	0.273	0.184
Away from home but not deployed	0.416	0.347	0.410	0.295
21–60 overtime days	0.246	0.224	0.218	0.246
61–120 overtime days	0.151	0.121	0.159	0.112
More than 120 overtime days	0.293	0.223	0.304	0.185
Much less time away than expected	0.079	0.090	0.133	0.166
Less time away than expected	0.102	0.104	0.145	0.152
More time away than expected	0.150	0.145	0.143	0.086
Much more time away than expected	0.131	0.125	0.098	0.046
Very poorly prepared for job	0.039	0.019	0.023	0.013
Poorly prepared for job	0.088	0.051	0.048	0.044
Well prepared for job	0.432	0.488	0.457	0.504
Very well prepared for job	0.264	0.284	0.335	0.292
AFQT category I	0.089	0.053	0.067	0.062
AFQT category II	0.401	0.344	0.439	0.458
AFQT category IIIA	0.274	0.259	0.257	0.268
AFQT category IIIB	0.206	0.313	0.212	0.178
Rural location	0.025	0.009	0.011	0.092
Micropolitan area	0.110	0.106	0.177	0.073
Metropolitan area	0.566	0.572	0.627	0.602
Some college	0.024	0.015	0.011	0.135
College graduate	0.058	0.031	0.008	0.016
Black	0.144	0.151	0.093	0.147
Hispanic	0.132	0.112	0.137	0.049
Other race	0.064	0.134	0.052	0.070
Married	0.461	0.370	0.335	0.441
Dual-service career spouse	0.021	0.017	0.006	0.090
Male	0.844	0.833	0.940	0.784
Infantry, gun crews, and seamanship	0.182	0.154	0.138	0.076
Electronic equipment repair	0.088	0.136	0.118	0.129

Table C.51—Continued

Statistic	Army	Navy	Marine Corps	Air Force
Communication and intelligence	0.150	0.076	0.065	0.090
Health care	0.110	0.121	0.000	0.065
Technical and allied	0.035	0.013	0.040	0.046
Functional support and administration	0.160	0.099	0.264	0.166
Electrical/mechanical equipment	0.131	0.272	0.200	0.296
Craftsworking	0.011	0.050	0.026	0.050
Service and supply handling	0.117	0.072	0.123	0.045
Nonoccupational	0.016	0.008	0.026	0.038
July 2002	0.047	0.035	0.033	0.042
March 2003	0.049	0.027	0.028	0.036
July 2003	0.108	0.113	0.113	0.103
November 2003	0.123	0.127	0.125	0.121
April 2004	0.110	0.115	0.116	0.112
August 2004	0.130	0.138	0.144	0.141
December 2004	0.109	0.114	0.113	0.110
March 2005	0.105	0.107	0.104	0.107
August 2005	0.109	0.109	0.111	0.111
December 2005	0.110	0.117	0.110	0.117
0–3 years of service	0.870	0.902	0.883	0.795
4–6 years of service	0.086	0.092	0.103	0.201
7–10 years of service	0.022	0.004	0.012	0.002
11 or more years of service	0.022	0.003	0.002	0.001
E-1	0.008	0.019	0.016	0.001
E-2	0.100	0.089	0.088	0.056
E-3	0.417	0.508	0.607	0.490
E-4	0.386	0.299	0.236	0.344
E-5	0.083	0.084	0.053	0.109

NOTE: Sample means were computed using survey weights.

Table C.52
Descriptive Statistics for Survey Sample, Second-Term-Plus Respondents

Statistic	Army	Navy	Marine Corps	Air Force
High work stress	0.517	0.492	0.451	0.495
High personal stress	0.457	0.390	0.411	0.359
Intention to stay	0.617	0.708	0.724	0.737
Observe reenlistment decision	0.567	0.625	0.734	0.665
Reenlisted (conditional on reenlistment decision being observed)	0.523	0.525	0.608	0.669
Nonhostile deployment only	0.135	0.130	0.162	0.124
Hostile deployment	0.359	0.218	0.244	0.211
Away from home but not deployed	0.485	0.493	0.556	0.506
21–60 overtime days	0.200	0.238	0.217	0.243
61–120 overtime days	0.172	0.151	0.172	0.173
More than 120 overtime days	0.418	0.273	0.373	0.294
Much less time away than expected	0.073	0.077	0.103	0.107
Less time away than expected	0.104	0.108	0.136	0.161
More time away than expected	0.174	0.137	0.126	0.110
Much more time away than expected	0.150	0.093	0.073	0.055
Very poorly prepared for job	0.013	0.007	0.008	0.009
Poorly prepared for job	0.037	0.027	0.025	0.043
Well prepared for job	0.448	0.467	0.430	0.469
Very well prepared for job	0.402	0.412	0.460	0.372
AFQT category I	0.036	0.052	0.038	0.053
AFQT category II	0.305	0.361	0.362	0.428
AFQT category IIIA	0.262	0.223	0.267	0.276
AFQT category IIIB	0.325	0.282	0.291	0.213
Rural location	0.026	0.013	0.013	0.052
Micropolitan area	0.108	0.105	0.154	0.064
Metropolitan area	0.654	0.706	0.703	0.642
Some college	0.117	0.039	0.039	0.849
College graduate	0.070	0.049	0.032	0.081
Black	0.306	0.191	0.218	0.198
Hispanic	0.109	0.106	0.148	0.049
Other race	0.061	0.120	0.049	0.056
Married	0.782	0.748	0.796	0.778
Dual-service career spouse	0.034	0.029	0.015	0.131
Male	0.878	0.889	0.947	0.824
Infantry, gun crews, and seamanship	0.223	0.085	0.155	0.079
Electronic equipment repair	0.064	0.166	0.095	0.096

Table C.52—Continued

Statistic	Army	Navy	Marine Corps	Air Force
Communication and intelligence	0.098	0.094	0.072	0.079
Health care	0.094	0.091	0.000	0.074
Technical and allied	0.037	0.021	0.036	0.039
Functional support and administration	0.215	0.167	0.313	0.286
Electrical/mechanical equipment	0.143	0.259	0.172	0.239
Craftsworking	0.014	0.048	0.020	0.051
Service and supply handling	0.111	0.070	0.111	0.055
Nonoccupational	0.002	0.000	0.026	0.001
July 2002	0.066	0.072	0.085	0.087
March 2003	0.072	0.073	0.077	0.076
July 2003	0.098	0.096	0.095	0.100
November 2003	0.109	0.110	0.108	0.110
April 2004	0.099	0.100	0.095	0.099
August 2004	0.136	0.135	0.131	0.129
December 2004	0.106	0.105	0.111	0.103
March 2005	0.092	0.094	0.093	0.093
August 2005	0.098	0.102	0.100	0.096
December 2005	0.105	0.104	0.104	0.099
0–3 years of service	0.088	0.015	0.033	0.013
4–6 years of service	0.237	0.245	0.284	0.177
7–10 years of service	0.224	0.226	0.262	0.223
11 or more years of service	0.452	0.514	0.421	0.588
E-3	0.015	0.016	0.023	0.004
E-4	0.192	0.111	0.088	0.076
E-5	0.265	0.341	0.324	0.398
E-6	0.280	0.333	0.298	0.283
E-7	0.169	0.133	0.158	0.188
E-8	0.057	0.044	0.072	0.034
E-9	0.019	0.020	0.034	0.016

NOTE: Sample means computed using survey weights.

Comparison with Hansen and Wenger's Navy Pay Elasticity

Hansen and Wenger (2002, 2005) focus on Navy zone A reenlistment from FY 1987 to FY 1999. They find that much of the difference in estimates of the effect of military compensation on reenlistment among past studies results from different regression specifications, regression models, and level of data aggregation. Their preferred specification is similar to ours with the notable exception that they control for grouped Navy occupations whereas we prefer occupation-specific fixed effects because bonuses are set by occupation. However, one of their robustness checks is a specification fixed effects by occupation (called "rating" in the Navy), which permits some comparison between their results and ours.

They define reenlistment as a new obligation of 36 months or more. Zone A refers to a service member in the first 21 to 60 months of active duty. Hansen and Wenger's pay variable is the annualized cost of leaving, and their regression model controls for pay grade, sea/shore rotation, whether voluntary separation incentive or special separation benefit was offered,[1] previous extension, length of service in months, race/ethnicity, married, number of children, age, AFQT, unemployment rate, geographic location, fiscal year, and rating fixed effects.

They estimate a logit regression and find a pay elasticity of 1.5 (Hansen and Wenger, 2002, p. 22). This estimate lies in the middle of the range of previous estimates (see the survey by Goldberg, 2001).[2] When they estimate the same model with fixed effects for individual ratings, the pay elasticity is 2.7 (p. 33). They compute the effect of a one-step increase in the reenlistment bonus, finding that it causes a 1.9-percentage-point increase in Navy zone A reenlistment when the model includes rating-group fixed effects and a 4.4-percentage-point increase when the model includes individual rating fixed effects (pp. 31, 33). It is important to note that, had they defined reenlistment as we do—namely, a new obligation of 24 months or more—their reenlistment rate would have included some service members who were not eligible to receive a bonus because they had not signed up to 36 months or more. Including these individuals would have had the effect of reducing their pay estimate and their bonus effect.

Using the rating-group specification, they also run separate models by year and compute the pay elasticity by year. The estimates are 1.4 to 1.6 for 1987–1992, 1.8–2.0 for the drawdown years 1993–1997, and 1.3–1.4 for 1998–1999. The simple average of the pay elasticity is

---

[1] However, they explain that zone A personnel were not eligible to receive voluntary separation incentive or special separation benefit, so their role in the regression is to indicate "a period during which the Navy's attitudes toward reenlistment were different" (Hansen and Wenger, 2002, p. 18).

[2] The results in Hansen and Wenger (2005), their subsequent journal article, are quite similar to those in the longer publication, Hansen and Wenger (2002). In this appendix, we refer to the earlier work because it contains more estimates that may be compared with our research.

1.45 in non-drawdown years and 1.90 in drawdown years, an increase of 31 percent. They state in the later journal article,

> Note that we estimate uniformly higher pay elasticities during the period approximating the military drawdown. It is not clear why estimated responsiveness to pay was higher during this time period; it is possible that this change is due to unobserved changes in the civilian job market or, more likely, a response to the dramatically changing landscape of the Navy. (Hansen and Wenger, 2005, p. 39)

The pay elasticity for the entire period is based on variation in the annualized cost of leaving (ACOL) over time and within rating groups, whereas the pay elasticity for a single year is based on variation within the rating group. Therefore, the pay elasticity for the entire period is not exactly comparable to the pay elasticity for any year. However, the elasticity for the entire period, 1.5, is near the simple average of the elasticity by year, which is 1.6. This suggests that much of the ACOL variation over the entire period comes from ACOL variation within rating groups rather than over time. Given that Hansen and Wenger's model controls for pay grade and years of service, this within-group variation probably comes mainly from bonus variation among the ratings in a rating group.

In contrast, when Hansen and Wenger shift to fixed effects for individual ratings, the ACOL variation probably comes mainly from bonus variation in a rating over time. This is the same variation we use. We argue that, because bonuses are set by individual occupation, the use of occupation-specific fixed effects reduces downward bias caused by bonus-setting behavior. The reduction in downward bias should be less if fixed effects by occupation group are used rather than fixed effects by individual occupation. In this light, we expect the Hansen and Wenger pay effect to be smaller when fixed effects by rating group are used, as opposed to fixed effects by rating, and that is what the results show.

In Appendix A, we further suggested that, after controlling for fixed effects for individual occupations, the bonus effect could be biased upward because of an omitted variable, such as the effort by career counselors to encourage sailors to reenlist in occupations offering a bonus. The same argument should apply to the ACOL variable, because much of the variation in ACOL probably traces to variation in bonus among the ratings in a rating group. This can help to account for Hansen and Wenger's higher pay elasticity and higher bonus effects in the drawdown years.

In summary, Hansen and Wenger (2002, 2005) obtain higher estimates of the pay effect when they use fixed effects for individual ratings rather than rating groups, and they obtain higher estimates for drawdown years. Their pay elasticity increases from 1.5 to 2.7 when fixed effects for individual ratings are used instead of fixed effects for rating groups, and their bonus effect increases from 1.9 percentage points to 4.4 percentage points. Their pay elasticity (using rating-group fixed effects) is 1.45 in non-drawdown years and 1.9 in drawdown years. Of these estimates, the one most directly comparable to our first-term bonus effect for the Navy is the 4.4-percentage-point increase, which compares with our estimate of 6.5 percentage points; both estimates come from models with fixed effects for individual ratings and derive primarily from bonus variation over time within the rating.[3] The estimates for drawdown versus

---

[3]   As mentioned, however, had Hansen and Wenger included new terms of 24–35 months for reenlistments, as we do, their pay effect would have been smaller, and the difference between their bonus estimate and ours would have been greater.

non-drawdown years are not as comparable because they are based primarily on bonus variation among ratings within rating groups for each year, but they suggest that responsiveness to bonuses was higher in drawdown years. This is consistent with the idea of an upward bias from an omitted variable in those years. If that bias were larger in our data for 1996–2007, it would have helped to account for our larger estimate of the bonus effect. The Navy drawdown years in our data are approximately 2003–2007.

References

Adler, Amy, Carl Castro, and Dennis McGurk, *Battlemind Psychological Debriefings*, U.S. Army Medical Research Unit–Europe, Walter Reed Army Institute of Research, Report No. 2007-001, 2007.

Ajzen, Icek, "From Intentions to Actions: A Theory of Planned Behavior," in Julius Kuhl and Jürgen Beckmann, eds., *Action Control: From Cognition to Behavior*, Berlin: Springer, 1985.

———, "The Theory of Planned Behavior," *Organizational Behavior and Human Decision Processes*, Vol. 50, No. 2, December 1991, pp. 179–211.

Campbell, Jason, Michael O'Hanlon, and Amy Unikewicz, "The State of Iraq: An Update," *New York Times*, December 22, 2007, p. A25.

Castaneda, Laura Werber, Margaret C. Harrell, Danielle M. Varda, Kimberly Curry Hall, Megan K. Beckett, and Stefanie Stern, *Deployment Experiences of Guard and Reserve Families: Implications for Support and Retention*, Santa Monica, Calif.: RAND Corporation, MG-645-OSD, 2008. As of April 6, 2009: http://www.rand.org/pubs/monographs/MG645/

Castro, Carl A., "10 Unpleasant Facts About Combat and What Leaders Can Do to Change Them," briefing, Walter Reed Army Institute of Research, January 15, 2008. As of April 6, 2009: https://www.battlemind.army.mil/assets/files/10_leaders_tough_facts_about_combat_brief.pdf

Defense Manpower Data Center, "Military Casualty Information," accessed for this research on October 20, 2008. As of October 20, 2008: http://siadapp.dmdc.osd.mil/personnel/CASUALTY/castop.htm

DMDC—*see* Defense Manpower Data Center.

Feldman, Noah, "Vanishing Act," *New York Times Magazine*, January 13, 2008, p. 11.

Filkins, Dexter, "Exiting Iraq, Petraeus Says Gains Are Fragile," *New York Times*, August 21, 2008, p. A6.

Fricker, Ronald D., *The Effects of Perstempo on Officer Retention in the U.S. Military*, Santa Monica, Calif.: RAND Corporation, MR-1556-OSD, 2002. As of April 6, 2009: http://www.rand.org/pubs/monograph_reports/MR1556/

Fricker, Ronald D., and Samuel E. Buttrey, *Assessing the Effects of Individual Augmentation (IA) on Active Component Navy Enlisted and Officer Retention*, Monterey, Calif.: Naval Postgraduate School, NPS-OR-08-003, August 2008. As of April 6, 2009: http://handle.dtic.mil/100.2/ADA485597

Goldberg, Matthew S., *A Survey of Enlisted Retention: Models and Findings*, Alexandria, Va.: Center for Naval Analyses, CRM D0004085.A2/Final, November 2001. As of April 6, 2009: http://www.cna.org/documents/D0004085.A2.pdf

Hansen, Michael L., and Jennie W. Wenger, *Why Do Pay Elasticity Estimates Differ?* Alexandria, Va.: Center for Naval Analyses, CRM D0005644.A2/Final, March 2002. As of April 6, 2009: http://www.cna.org/documents/D0005644.A2.pdf

———, "Is the Pay Responsiveness of Enlisted Personnel Decreasing?" *Defence and Peace Economics*, Vol. 16, No. 1, 2005, pp. 29–43.

Hattiangadi, Anita U., Deena Ackerman, Theresa H. Kimble, and Aline O. Quester, *Cost-Benefit Analysis of Lump Sum Bonuses for Zone A, Zone B, and Zone C Reenlistments: Final Report*, Alexandria, Va.: Center for Naval Analyses, CRM D0009652.A4/1Rev, May 2004. As of April 6, 2009:
http://www.cna.org/documents/D0009652.A4.pdf

Hoge, Charles W., Jennifer L. Auchterlonie, and Charles S. Milliken, "Mental Health Problems, Use of Mental Health Services, and Attrition from Military Service After Returning from Deployment to Iraq or Afghanistan," *Journal of the American Medical Association*, Vol. 295, No. 9, March 1, 2006, pp. 1023–1032.

Hosek, James, Jennifer Kavanagh, and Laura Miller, *How Deployments Affect Service Members*, Santa Monica, Calif.: RAND Corporation, MG-432-RC, 2006. As of April 6, 2009:
http://www.rand.org/pubs/monographs/MG432/

Hosek, James, and Christine E. Peterson, *Reenlistment Bonuses and Retention Behavior*, Santa Monica, Calif.: RAND Corporation, R-3199-MIL, 1985. As of April 6, 2009:
http://www.rand.org/pubs/reports/R3199/

Hosek, James, and Mark Totten, *Does Perstempo Hurt Reenlistment? The Effect or Long or Hostile Perstempo on Reenlistment*, Santa Monica, Calif.: RAND Corporation MR-990-OSD, 1998. As of April 6, 2009:
http://www.rand.org/pubs/monograph_reports/MR990/

———, *Serving Away from Home: How Deployments Influence Reenlistment*, Santa Monica, Calif.: RAND Corporation, MR-1594-OSD, 2002. As of April 6, 2009:
http://www.rand.org/pubs/monograph_reports/MR1594/

Hurd, Michael, "Mortality Risk and Bequests," *Econometrica*, Vol. 57, No. 4, July 1989, pp. 779–813.

Karney, Benjamin R., and John S. Crown, *Families Under Stress: An Assessment of Data, Theory, and Research on Marriage and Divorce in the Military*, Santa Monica, Calif.: RAND Corporation, MG-599-OSD, 2007. As of April 6, 2009:
http://www.rand.org/pubs/monographs/MG599/

Lyle, David S., "Using Military Deployments and Job Assignments to Estimate the Effect of Parental Absences and Household Relocations on Children's Academic Achievement," *Journal of Labor Economics*, Vol. 24, No. 2, April 2006, pp. 319–350.

MacDermid, Shelley M., Rita Samper, Rona Schwarz, Jacob Nishida, and Dan Nyaronga, *Understanding and Promoting Resilience in Military Families*, West Lafayette, Ind.: Military Family Research Institute at Purdue University, July 2008.

McClintock, Jean, "Predeployment Briefing," Army Community Service, Ft. Carson, Colo., undated. As of October 13, 2008:
http://community.carson.army.mil/ACS/pdf/predeployment.html

Military OneSource, homepage, accessed for this research on October 13, 2008. As of April 6, 2009:
http://www.militaryonesource.com/

Office of the Under Secretary of Defense (Comptroller), Chapter 14, "Special Pay or Bonus—Qualified Members Extending Duty at Designated Locations Overseas," Vol. 7A, "Military Pay Policies and Procedures—Active Duty and Reserve Pay," *DoD Financial Management Regulation*, September 2008. As of April 6, 2009:
http://www.defenselink.mil/comptroller/fmr/07a/index.html

———, Chapter 10, "Special Pay—Duty Subject to Hostile Fire or Imminent Danger," Vol. 7A, "Military Pay Policies and Procedures—Active Duty and Reserve Pay," *DoD Financial Management Regulation*, May 2009. As of July 27, 2009:
http://www.defenselink.mil/comptroller/fmr/07a/index.html

Quester, Aline O., Anita U. Hattiangadi, Lewis G. Lee, and Robert W. Shuford, *Marine Corps Deployment Tempo and Retention in FY05*, Alexandria, Va.: Center for Naval Analyses, CRM D0013786.A2/Final, March 2006. As of April 6, 2009:
http://www.cna.org/documents/D0013786.A2.pdf

Quester, Aline O., Anita U. Hattiangadi, and Robert W. Shuford, *Marine Corps Retention in the Post-9/11 Era: The Effects of Deployment Tempo on Marines with and Without Dependents*, Alexandria, Va.: Center for Naval Analyses, CRM D00113462.A1/Final, January 2006. As of April 6, 2009:
http://www.cna.org/documents/D0013462.A1.pdf

Rinehart, Michael, "Pre-Deployment Battlemind for Warriors (Preparing for War: What Warriors Should Know and Do)," briefing, PSB04001/1, May 22, 2008. As of April 6, 2009:
https://www.battlemind.army.mil/assets/files/pre_deployment_battlemind_for_warriors.pdf

Savych, Bogdan, *Effects of Deployments on Spouses of Military Personnel*, Santa Monica, Calif.: RAND Corporation, RGSD-233, 2008. As of April 6, 2009:
http://www.rand.org/pubs/rgs_dissertations/RGSD233/

Simon, Curtis, and John Warner, "Deployment, Reenlistment Bonuses, and Reenlistment: An Analysis of Army Reenlistment Using Data from FY 2002–2006," unpublished manuscript, April 2009.

Tanielian, Terri, and Lisa H. Jaycox, eds., *Invisible Wounds of War: Psychological and Cognitive Injuries, Their Consequences, and Services to Assist Recovery*, Santa Monica, RAND Corporation, MG-720-CCF, 2008. As of April 6, 2009:
http://www.rand.org/pubs/monographs/MG720/

Vanden Brook, Tom, "DoD Data: More Forced to Stay in Army," *USA Today*, April 23, 2008. As of May 29, 2009:
http://www.usatoday.com/news/military/2008-04-21-stoploss_N.htm